GREAT DISCOVERIES IN SCIENCE

X-rays

Kristin Thiel

New York

Published in 2018 by Cavendish Square Publishing, LLC
243 5th Avenue, Suite 136, New York, NY 10016

First Edition

Website: cavendishsq.com

CPSIA Compliance Information: Batch #CS17CSQ

Library of Congress Cataloging-in-Publication Data

Names: Thiel, Kristin.
Title: X-rays / Kristin Thiel.
Description: New York : Cavendish Square, 2018. | Series: Great discoveries in science| Includes index.
Identifiers: ISBN 9781502627780 (library bound) | ISBN 9781502627797 (ebook)
Subjects: LCSH: X-rays--Juvenile literature. | Radiography--Juvenile literature.
Classification: LCC QC481.T45 2018 | DDC 616.07'572--dc23

Editorial Director: David McNamara
Editor: Caitlyn Miller
Copy Editor: Michele Suchomel-Casey
Associate Art Director: Amy Greenan
Designer: Lindsey Auten
Production Coordinator: Karol Szymczuk
Photo Research: J8 Media

The photographs in this book are used by permission and through the courtesy of: Cover, p. 8 Puwadol Jaturawutthichai/Shutterstock.com; p. 4 George Frey/Bloomberg/Getty Images; p. 12 Imagno/Hulton Archive/Getty Images; p. 14 Science Source; p. 17 Susan Watts/NY Daily News Archive/Getty Images; pp. 19, 31 Photo Researchers/Alamy Stock Photo; p. 23 John Reader/Science Photo Library/Getty Images; p. 26 CNRI/Science Photo Library/Getty Images; p. 28 Oxford Science Archive/Print Collector/Getty Images; pp. 35, 71 SSPL/Getty Images; p. 37 Chronicle/Alamy Stock Photo; p. 39 Fototeca Gilardi/Hulton Archive/Getty Images; p. 42 Science Source/Getty Images; p. 46 Paul Cannon/AP Images; p. 48 ImageBROKER/Alamy Stock Photo; p. 51 Hulton Archive/Getty Images; p. 58 Yakoniva/Alamy Stock Photo; p. 60 Edwin J. Houston/Wikimedia Commons/File:Fluoroscopy procedure 1909.jpg/Public Domain; p. 63 NYPL/Science Photo Library/Getty Images; p. 67 Designua/Shutterstock.com; p. 78 John Parrot/Stocktrek Images/Getty Images; p. 80 NASA/JPL-Caltech/ESA/CXC/STScI/Wikimedia Commons/File:Center of the Milky Way Galaxy IV – Composite.jpg/Public Domain; p. 83 Wellcome Images/Wikimedia Commons/File:Colloid cancer involving the toe Wellcome L0062410.jpg/CC BY 4.0; p. 88 Paul Sakuma/AP Images; p. 91 Chrisi1964/Wikimedia Commons/File:Archaeopteryx im Jura-Museum Eichstätt.jpg/CC BY 4.0; p. 97 Mariakraynova/Shutterstock.com; p. 101 Cordelia Molloy/Science Source.

Printed in the United States of America

Contents

A Transportation Security Administration checkpoint at Salt Lake City International Airport X-rays passengers for weapons.

Introduction: Changing How We See the World

From its humble beginnings as an accidental discovery, the X-ray has come a long way. In 2009, a Science Museum of London poll named the X-ray the most important modern scientific discovery over penicillin, various rocketry, one of the first computers, the first motorized car, and the electric telegraph. It even beat the DNA double helix, which came in number three in the poll of nearly fifty thousand votes. It seems fitting that X-rays took precedence, considering the double helix was found in large part because of Wilhelm Röntgen's discovery of the rays!

In 2010, the 115th anniversary of Röntgen's work with X-rays was commemorated with a Google Doodle. The search engine's name was depicted as a **radiograph**, and bones and foreign objects, like a "swallowed" key, were visible within the ghostly shadows of "tissue."

The X-ray made itself known to humans through a splashy demonstration for Röntgen, the well-respected but definitely not flashy physics professor in a midsize German city. Röntgen realized that X-rays existed when he saw the bones of his own hand appear on a screen in his laboratory one dark November night in 1895. He wasn't expecting X-rays, but he would be glad they'd shown up—they earned him lasting fame and professional accolades, including a Nobel Prize. Most importantly, they changed our world. Even Röntgen focused on the discovery's applications and not on his individual role in it. He was pleased to be acknowledged for his work, but he refused any major claim on X-rays, saying that such a discovery should not be patented by one but "owned" by everyone. He may have even found it odd that we call it a discovery, since that places the emphasis on the human. X-rays have always existed—it's more accurate to say that humans finally came to be able to see them.

Nearly immediately after Röntgen's December 28, 1895, announcement of his discovery and the publication of his initial findings, complete with the first two **shadowgraphs**, scientists and the general public alike found use for the new technology. Surgeons saved more lives on the battlefield by using X-rays to locate bullets, rather than poking their dirty fingers into wounds. In the hospital, they diagnosed kidney stones, saw early signs of tuberculosis, and located the breaks in painful bones. Dentists spotted cavities and cancers. For the first time, researchers saw DNA, life's instruction manual. X-ray machines found less useful applications, too. They became a frivolity at parties and carnivals, and people chose which shoes they were going to buy after wiggling their toes in a few different pairs in department store **fluoroscopes**.

In Röntgen's time, scientists, medical professionals, and the general public welcomed X-rays with open arms—even while admitting that what X-rays were capable of doing was undeniably weird! On January 25, 1896, less than a month after Röntgen's announcement that he'd discovered X-rays, the British humor magazine *Punch* published a poem directed at the scientist: "O, Röntgen, then the news is true," it began and quickly poked fun at the "grim and graveyard humour" of his discovery. "We do not want, like Dr. Swift, / To take our flesh off and to pose in / Our bones, or show each little rift / And joint for you to poke your nose in."

Yet two years after Röntgen made X-rays public, their health risks came to the fore. Pioneers in X-ray research lost hair, their skin blistered, and the pain in their joints became so intense that some had to experience amputation of hands or arms. Many died of cancer. Today, the International Agency for Research on Cancer (IARC), which is a part of the World Health Organization (WHO), classifies X-ray radiation as a "known human **carcinogen**."

Still, over the years, X-rays have become even more commonplace. If you go to the dentist, the hygienist takes X-rays of your mouth. If a baby swallows something she shouldn't have been playing with, she's destined for an X-ray at the emergency room. If you fly to visit a relative, you must stand in an X-ray machine, arms and legs wide, while your backpack goes through its own X-ray next to you. We continue to use this awesome technology—with standards and precautions in place to try to mitigate the health risks—because the benefits are so very great. From the early diagnosis and successful treatment of cancer, to the safe examination of delicate artifacts thousands of years old, to the realization of celestial bodies many light-years away, X-rays have shaped our modern world.

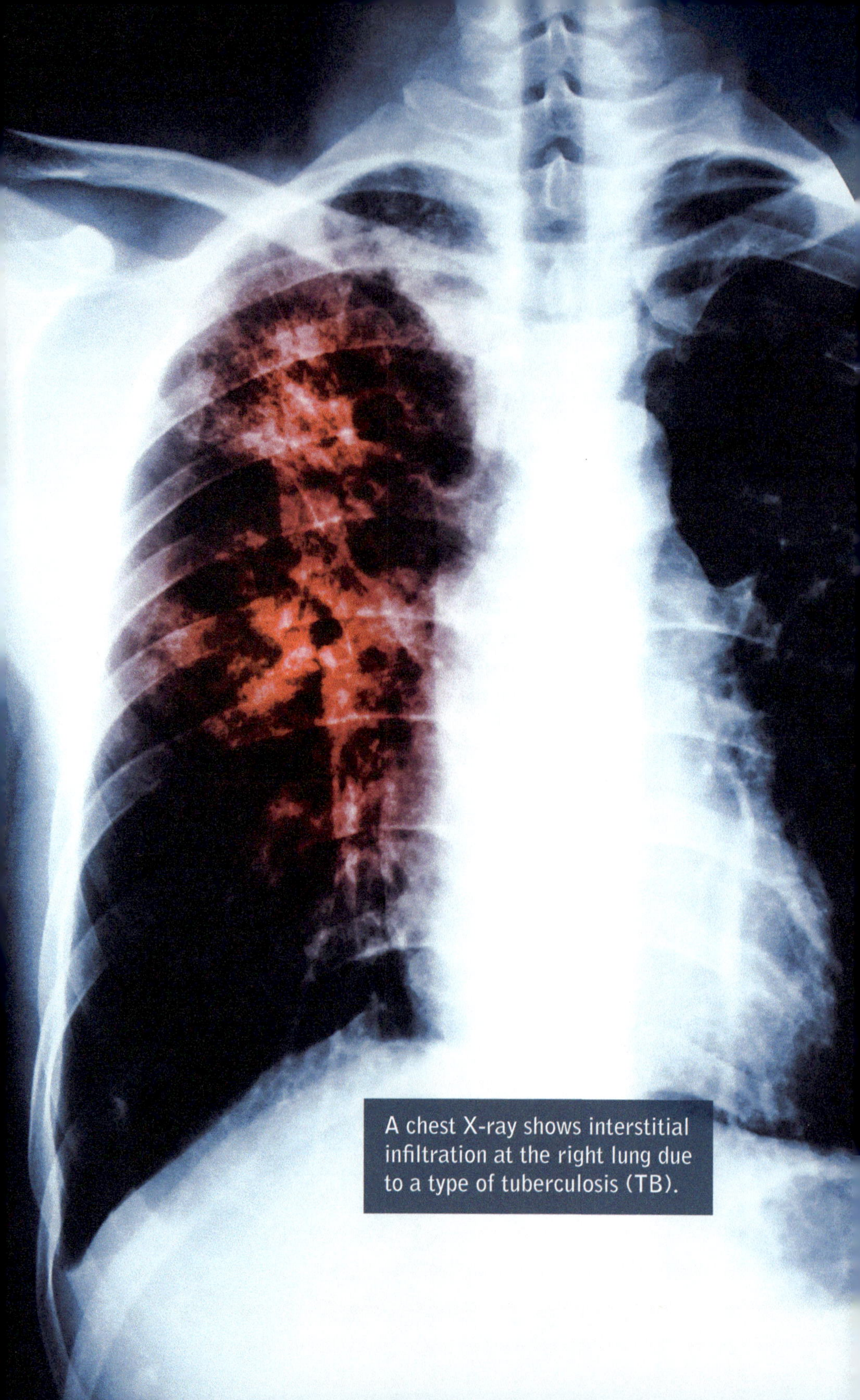

A chest X-ray shows interstitial infiltration at the right lung due to a type of tuberculosis (TB).

CHAPTER 1

What's Inside?

Once X-rays became commonplace, some people said the images they produced showed the truth of the world. The beams did, after all, get to the core of a person, and radiographs showed those unmasked insides to anyone who cared to look. Though this was a spiritual, or at least poetic, view of what X-rays did, it is true even from scientific and medical perspectives that X-rays show a truth previously unavailable to us.

Particularly when discussing health concerns, it's important to be clear that X-rays are not foolproof; they're not magic. There is no magic solution to **disease**. X-rays don't save everyone; no technology or treatment can. But they help—a lot. Before this technology, dentists, surgeons, and doctors made many medical missteps simply because they didn't have a clear picture of what was happening inside of their patients. Treating patients before X-rays was a little like driving a car with your eyes closed. With your experience and use of your other senses, you can do a fair job—if luck is also on your side and no one steps in front of your car, the road doesn't curve, or the light doesn't turn red. If your eyes are open, you can prevent accidents and address problems much more successfully. With the "sight" of X-rays, medical professionals diagnosed patients earlier and treated them more successfully; before X-rays, patients stayed in pain. Today, those same patients would have found relief, and many may not have died.

The TRUTH of the TOOTH

Dentistry is an ancient practice—the record of "tooth worms," a description for dental decay that is both whimsical and disturbing, dates from 5000 BCE. Through archaeological research, we know that dentists from long ago used ingenious and successful techniques, especially considering the limited technology they were working with. Still, without X-rays to show what was happening beyond their range of vision, so much remained unknown to them, and big medical problems happened because of unseen or misunderstood problems in the mouth.

Types of Dental X-rays

Dental X-rays show problems in the whole mouth, including the parts that cannot be seen by the naked human eye, such as between teeth and under gums. Two types of dental X-rays focus on teeth: bitewing X-rays allow dental professionals to check for decay between teeth and bone loss in the bones that support the teeth. Periapical X-rays take images of problems below the gums and in the jaw, including tumors and bone changes symptomatic of other diseases. Two other types of dental X-rays take a broader view of the mouth: occlusal X-rays help dentists examine the roof and floor of the mouth, and panoramic X-rays even look into the sinuses, nasal area, and jaw joints.

Early Dental Treatments

It's important to note that medical professionals who lived well before the discovery of X-rays took diagnostic and treatment steps that professionals today take after reviewing the information provided by X-rays. They just did so without the guidance and assurance that X-rays provide today. Hippocrates and Aristotle wrote in 500–300 BCE Greece about treating decayed teeth and gum disease, which bitewing and periapical X-rays today help to stop to a great degree. Celsus, a Roman medical writer in 100 BCE, documented jaw fractures, again often confirmed today by occlusional X-rays. These confirmations help assure dentists and patients of the true source of the pain, where exactly along the jaw the break is located, and what the best treatment is. Occlusional X-rays also help to show cleft palates; however, the first known cleft lip repair happened during the 255–206 BCE Ch'in dynasty of China. The Guild of Barbers was established in France in 1210 CE, which included a class of surgeons formally educated and trained to perform complex surgical operations—all without the help of X-rays in plotting out the necessary steps of each surgery. (Hair-cutting barbers were some of the world's first surgeons because they were skilled with knives!) By the 1600s, Chinese dental surgeons existed and dealt with tonsillar abscesses (pus behind the tonsils showing infection) and eptheliomas of the lip (growths deforming the lip).

The Problems with Early Dental Treatments

X-rays allow dentists to catch more issues more often and more quickly than dentists working without that aid. The ingenuity and common sense of early dentists were simply not enough for all conditions all the time—some issues of the mouth even ended up killing the patients. For years during the

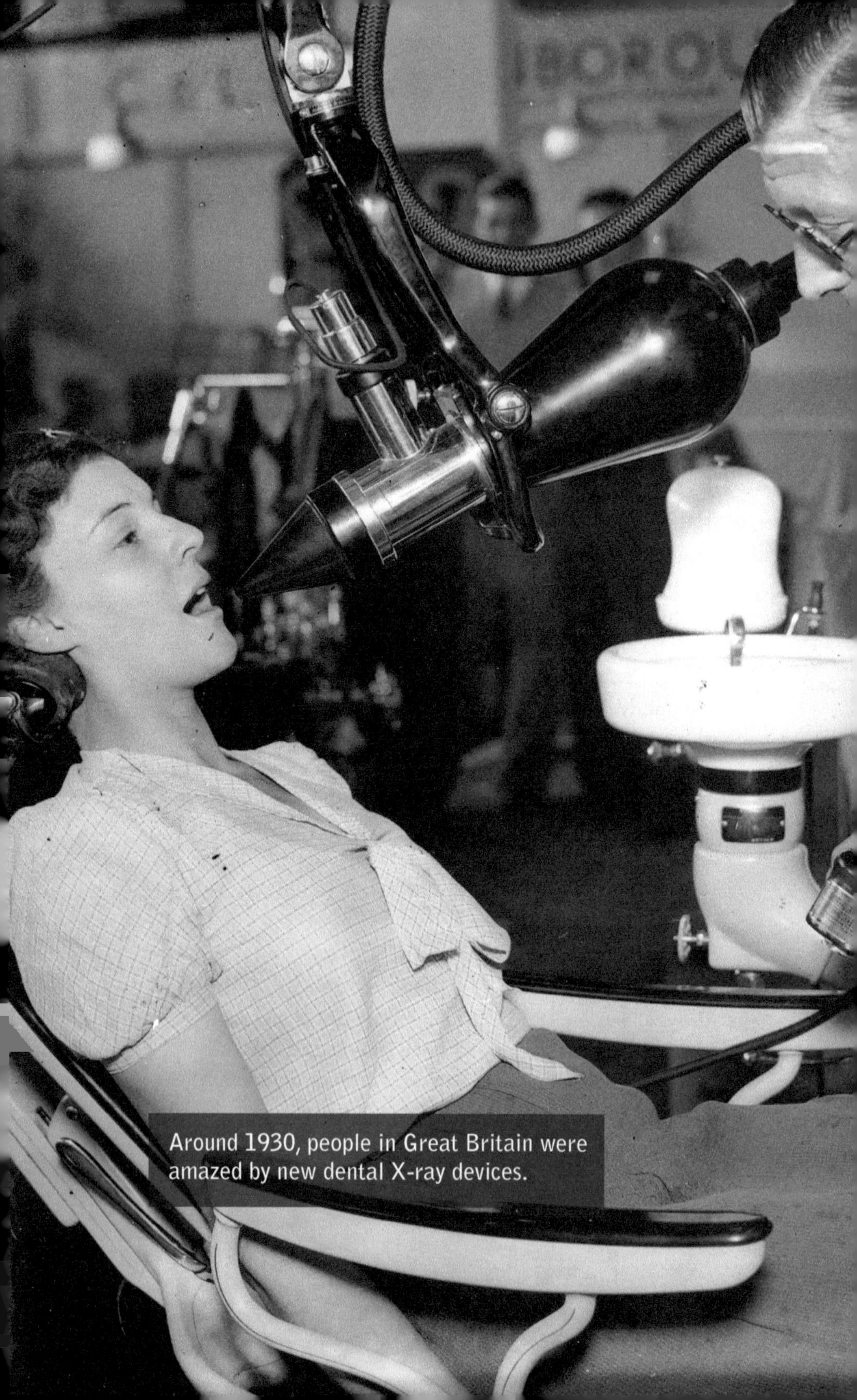

Around 1930, people in Great Britain were amazed by new dental X-ray devices.

1600s, the London Bills of Mortality listed "Teeth" as the fifth or sixth most common cause of death. The first surgery to use **anesthetic** to numb the patient didn't happen until 1846, so until then, only patients suffering the most serious emergencies went under the knife. If dentists could have at least X-rayed their patients, more might have risked the pain of surgery. However, without the definitive proof of injury or illness that X-rays later would provide, dentists would have been hesitant to suggest invasive procedures—and patients would have been reticent to accept the excruciating pain. Surgery was a last desperate step taken in only the worst situations, not only because of the lack of anesthetic but because no one knew what exactly was wrong.

STICKS and STONES

Before X-rays were discovered and medical X-ray machines were invented, broken bones, bullets, and other issues hidden within the body were all diagnosed by a doctor guessing while poking around the exterior of a patient's body. Often, it wasn't even a trained doctor who was doing the prodding—in the 1700s, for example, most people couldn't afford to see someone educated in medicine. Instead, they went to the neighborhood bonesetter, a person who had completed an apprenticeship and learned how to evaluate a break, set a broken bone back into place, and immobilize it to heal.

In his 1769 book *Domestic Medicine*, physician William Buchan described how to determine if a bone was broken. He first looked, of course, for the fractured edge of a bone sticking through the skin. If there was no sign of that, he checked to see if the hurting limb appeared shortened or otherwise changed in form. If nothing out of the ordinary was apparent by sight, he grasped the limb and moved it around, listening for the grating sound of broken pieces scraping against each other.

This ninth-century Greek surgical manuscript shows surgeons using the Hippocratic treatment of joint dislocation.

If the bone was broken, the pieces were pushed back together—again, there was nothing to numb the pain, outside of the patient taking a swig of alcohol. The pieces were held immobile not with a cast but with a splint of wood, leather, or pasteboard (glue enough sheets of paper together and you have a sturdy, if not waterproof, board).

While excruciatingly painful, this whole process wasn't terribly complicated, assuming the break was relatively clean. But consider snapping a pretzel rod in half—does it usually break in two pieces, or do flecks of pretzel go flying? When a bone breaks, the same thing happens. Even if the bone appears to break in only two pieces, chips of it can flake off and lodge in the body. Left there, loose fragments of shattered bone will cause infection. Buchan suggested immediate amputation if the bonesetter was at all worried about lost pieces of bone. There was no time to wait and see because before even three days had passed, the infection could be so bad that not even amputation could save the patient from death. Today, X-ray images can show even the tiniest of hidden bone fragments within the body.

Modern Bonesetters

We can learn from modern practitioners what bonesetters through the ages have done without X-ray technology to aid them. Investigators for a 2011 study on bonesetters in Nigeria reported the process these professionals follow. First, the bonesetter determines whether a break is closed or open, also called compound fracture. In the latter, the broken bone breaks through the skin. If the break is open, the patient must go to a clinic to have the wound cleaned and closed. The patient then returns to the bonesetter with what is effectively now a closed fracture.

Second, a bonesetter identifies the exact location of the break by physically examining the limb and thumping fingers

Bonesetters in Popular Culture

As a part of our past, bonesetters are an important chapter in our medical history. And in some parts of the world, bonesetters still carry out their work treating patients. Most don't have any formal medical training but have learned through experience and practice to locate breaks in bones, set them, and wrap them.

Stories of bonesetters in popular culture highlight our fascination with their work. An orthopedic surgeon and instructor in **orthopedics** at the University of North Carolina Medical School has written a historical fiction series starring a young bonesetter. The boy, living in the caveman era, learned to set bones, which set him on the path to becoming a leader in his society. The bestselling author Amy Tan wrote *The Bonesetter's Daughter* (2001), in which the main characters are descendants of a bonesetter in China. The book was even adapted into an opera, the lyrics of which were also written by Tan, which the San Francisco Opera premiered in 2008.

Amy Tan wrote the novel *The Bonesetter's Daughter.*

along the skin to feel and hear differences. Third, once the fracture is found, the bonesetter manipulates the limb to realign the pieces of bone. Fourth, an herbal cream, "Ufie," and an incantation soothe the traumatized limb, which is then splinted with cloth, hard cardboard, or plywood.

DEATH by BULLET and DOCTOR

James A. Garfield, the twentieth president of the United States and the second of four assassinated presidents in the country's history, is one of the most famous examples of when a life might have been saved if only the medical X-ray existed at the time. Garfield had been on the job for less than four months when he was assassinated. A delusional citizen, upset that he hadn't been given the diplomatic title of US minister to France and operating under the belief that God had told him to kill the president, walked up to Garfield where he stood waiting for a train and shot him.

Getting Shot Wasn't a Death Sentence

Neither the bullet that hit his arm nor the one that lodged in his back immediately killed Garfield when he was shot on July 2, 1881. Rather, the medical protocol and technology of the day killed him—after eighty long days, on September 19 of that year. Doctors today who have examined the records of his injuries and treatment say his wounds would have been nonlethal if they'd been sustained even a few years later. Unfortunately, Garfield was living right before great leaps in medical innovation were taken.

The first bullet grazed his arm. The second, which buried itself in his back, was one of concern, but the situation wasn't dire. It pierced his vertebra but missed his spinal cord, settling

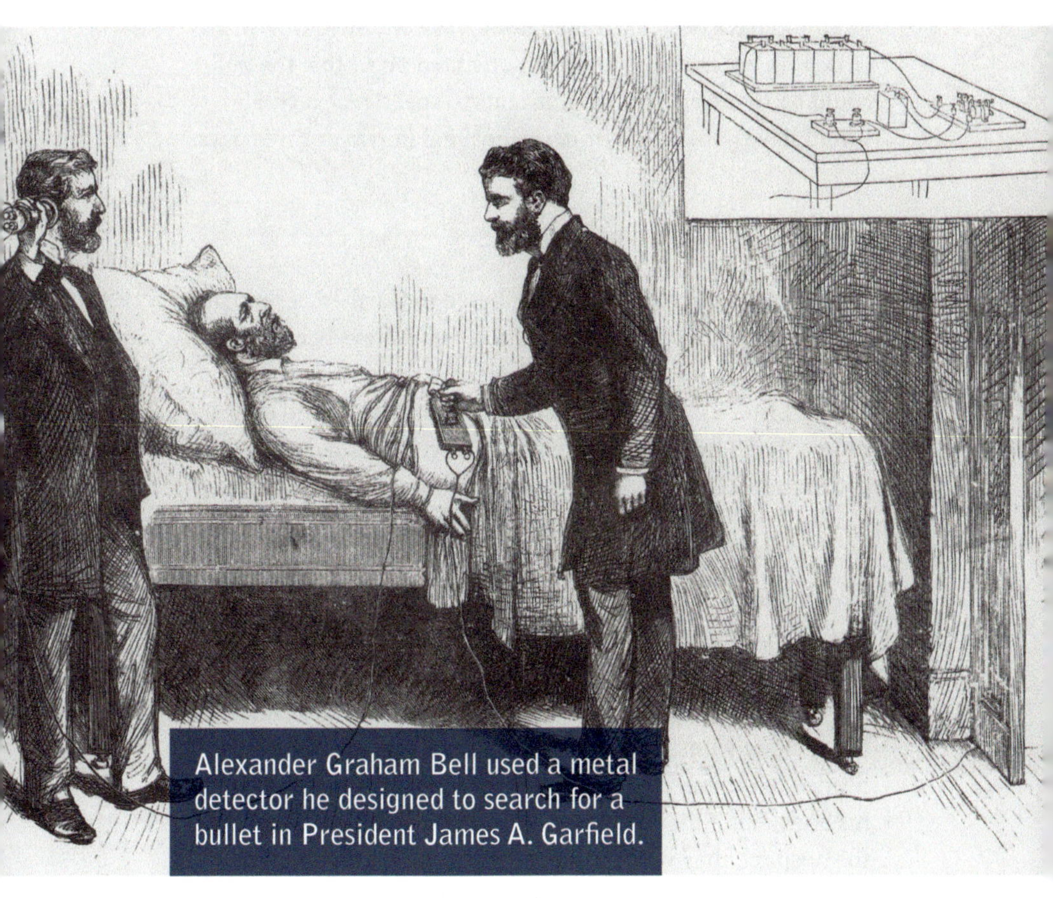

Alexander Graham Bell used a metal detector he designed to search for a bullet in President James A. Garfield.

above his pancreas in adipose tissue, which stores energy in the form of fat and cushions and insulates the body. In other words, if you're going to be shot in the back, that is not a bad place. Today, Garfield would have been rushed to a hospital where he would have been examined noninvasively, including with X-rays, in a sterile environment; operated on so that the bullet could be removed; and fed nutrients to speed recovery. He would have gone home from the hospital in two or three days.

Faulty Treatment

Though a variety of factors contributed to Garfield's eventual death, none would have played a role had X-rays been available.

By 1881, a lot of Europe had accepted British surgeon Joseph Lister's ideas about keeping hospital wards and operating rooms sterile, which he'd proved was effective at reducing patient infection and death. American doctors on the whole, however, hadn't yet accepted that **germs** caused disease, let along existed. They still thought the longstanding **miasma theory** had merit—that bad air caused disease or even was disease. Garfield fell to the train station platform, and every doctor who happened to be nearby rushed to his side and began examining him, inadvertently transferring dirt and germs from the decidedly unsterile environment into his wounds. They needed to quickly understand where the bullets had lodged in his body, but if they'd known an X-ray machine was waiting at the hospital, they wouldn't have been so quick to touch his open wounds with bare hands at a public transit station.

Multiple doctors continued to transfer germs to the vulnerable patient even at the hospital, which was hardly cleaner than the outside world. Doctors didn't wash their hands, didn't wear gloves and gowns, and didn't sterilize their medical instruments. And because they feared the bullet had pierced Garfield's intestines, they severely limited his food and the way he ingested it. He was fed a liquid diet of beef broth, egg yolks, milk, whiskey, and opium. He lost over 100 pounds (45 kilograms). If they could have seen inside him with X-rays and known his digestive track was free and clear, they wouldn't have hastened his death with starvation.

Medical professionals knew the key to helping their patient was to figure out the path the bullet took through his body and where it had stopped. They would have loved what X-rays could have done for them. Instead, they enlisted the help they did know about. A bullet is made of metal, and Alexander Graham Bell, the inventor of the telephone, had been developing a kind of metal detector that he called an induction balance. Doctors asked Bell to wave his device over the president. His attempts failed to locate the bullet. Historians suspect this was because he was lying on a mattress with metal springs, rare in that era, and because the self-appointed lead physician refused to allow Bell to search anywhere but over the right side of Garfield's body because that was where he was sure the bullet lay.

Infection and starvation killed the president. X-rays could have helped reduce the likelihood of both. Knowing where the bullet was, doctors wouldn't have had to examine him with their dirty fingers as much as they did, and they wouldn't have feared feeding him. Of course, considering the lack of cleanliness in hospitals at the time, infection caused by dirty surgical instruments and unwashed hands during postsurgical exams might have killed him anyway. (One could also argue that the ultimate prevention of death would have been a prevention of the shooting: even after President Abraham Lincoln's assassination sixteen years before, there was still no Secret Service.)

CANCER

Cancer is often thought of as a modern disease, and in a way, it is. It's likely that, without modern ills like tobacco usage, unhealthy diets, pollution from industrialization, and stress, which are linked to cancer, more people today would contract cancer simply because we're living longer than humans ever have. We're living long enough for cancer to have a chance to latch on.

Early Records of Cancer

Paleopathologists examining the remains of people who died thousands of years ago have found evidence of disease, including cancer, in ancient bones. For example, in the early 2000s, deep in a cave in Russia, archaeologists found the bones of a couple who had lived 2,700 years ago—and the archaeologists' paleopathologist colleagues found evidence in bone scars of the malignant cells of the prostate cancer that lived inside the man.

Written records exist, as well. Dating from 3000 BCE, the Edwin Smith Papyrus is the first known medical textbook that mentions cancer, although not by that term. Named for the Egyptologist Edwin Smith, who bought the scrolls in 1862, the text consists of forty-eight cases—including symptoms, diagnosis, prognosis, treatment, and context or follow-up—that demonstrate ancient Egyptians' vast knowledge about health. Eight of those forty-eight cases discuss tumors or ulcers of the breast, which doctors targeted with a "fire drill." This hot metal tool cauterized, or burned, the infected area to destroy the cancerous tissue. Ancient physicians concluded that there was no treatment for the cause of those symptoms. Writings from Egyptian physician Imhotep, who lived around 2600 BCE, noticed a seemingly untreatable illness with the symptom of a "bulging mass" in the breast. And ancient Greeks knew that a mastectomy would help a woman with such a symptom.

Hippocrates was a physician living in Greece in 460–370 BCE who would become known as the Father of Medicine. He was the first to use the label "cancer" to describe non-ulcer-forming and ulcer-forming tumors. *Carcinos* and

Neanderthal man *Homo neanderthalensis*'s skull and bones were found in 1908 in France.

carcinoma, the words he used, mean "crab" in Greek, and, indeed, cancers can spread with projections that look like the shape of a crab. The Greek Galen (130–200 CE), a physician in Rome, described tumors with the word *oncos* (Greek for "swelling"), indicating an understanding of how they grow. Another Roman physician, Celsus (28–50 BCE), translated the Greek term into Latin's word for "crab," *cancer*.

As of 2010, researchers had recorded about two hundred different prehistoric cancer patients. Part of the reason for the low number is because, again, likely fewer people then had cancer compared to now; also, researchers have tested only a small number of archaeological remains.

Breast Cancer

The proof remains that cancer did exist, which means that early diagnosis (along with, of course, better treatment options) would have helped a lot of people. Breast cancer is a type of cancer that X-rays help doctors catch early. Doctors advocated for radical mastectomies—the removal of the entire breast, including the muscle underneath and the lymph nodes—from 1882, when Johns Hopkins surgeon William S. Halsted performed the first one, through the 1960s. This was because by the time they caught the disease, it was too far along to treat any other way. Breast cancer ultimately displays itself prominently, with lumps that one can see or feel even with exterior physical examination. But if it can be found before then, it's much easier to treat, with a higher success rate and less likelihood of spreading to other parts of the body.

Bone Cancer

Bone cancer is another cancer that can be seen in X-rays, though a complete diagnosis can't be made until a biopsy is also completed. That's because other diseases, such as bone infections, can present in a very similar fashion to bone cancer. Even then, X-rays can still help: while performing a CT-guided needle biopsy, a **radiologist** slowly approaches a suspicious mass with a biopsy needle, taking **CT scans** until one shows the needle has made contact with its intended target.

Multiple kinds of X-ray technology can be used to detect different kinds of bone cancer or the cancers that result when bone cancer spreads to other parts of the body. A basic X-ray can show the ragged look of a bone that is supposed to be smooth; it can also show holes or tumors that extend from the bone into the surrounding muscle or fat. X-rays can even give radiologists an idea if a tumor is malignant or benign because the two look different!

A chest scan can show if the bone cancer has traveled to the lungs. A CT scan is also interested in the spread of bone cancer. They're able to see defects in lots of organs, including the lungs and liver, as well as in lymph nodes. The use of X-rays for detecting bone cancer provides a clear example of how far-reaching an X-ray's help extends. With this technology, a pain in the arm could be diagnosed, and a future pain in the chest could be prevented.

TB

Tuberculosis (TB) was a huge killer for a long time in the United States. Around the time the X-ray was invented, 450 Americans a day, most young, between fifteen and forty-four years old, wasted away from fatigue, punctuated by the persistent coughing of thick white phlegm and blood. The *Mycobacterium tuberculosis* **bacterium** spread through crowded cities along the spittle of coughing TB patients. Those who survived an initial bout usually experienced recurrences for the rest of their lives. The American Lung Association formed in 1904 to help beat the disease. Until the development of antibiotics in the mid-twentieth century, there was no treatment for TB except rest.

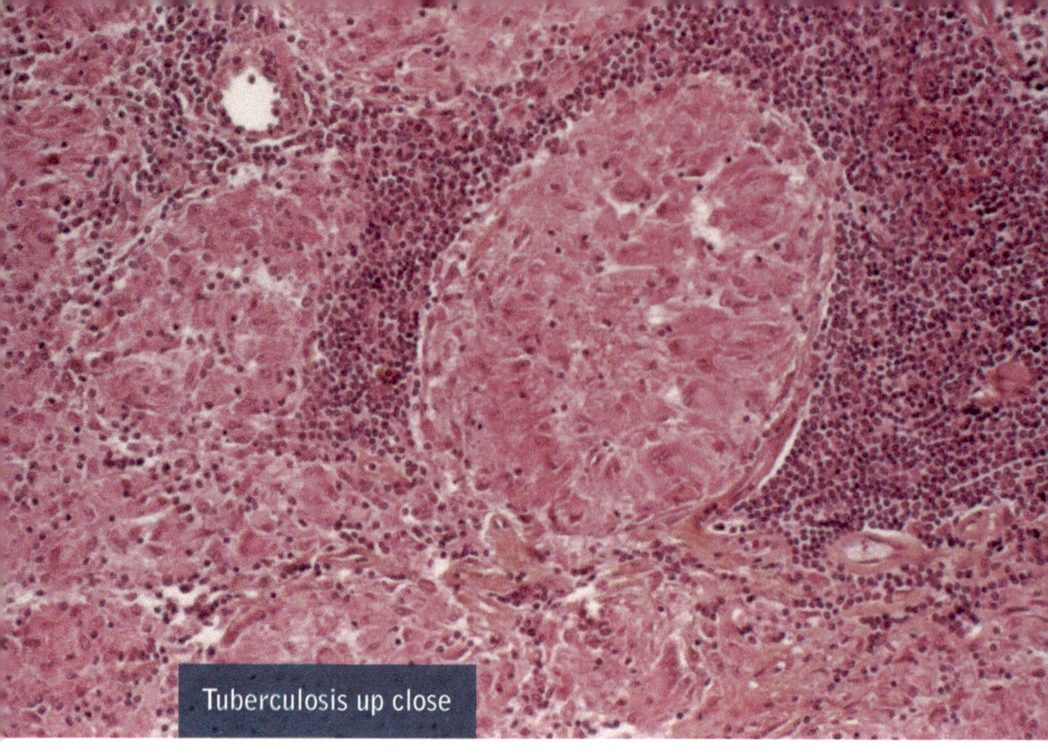
Tuberculosis up close

And until the X-ray, diagnosis of TB was attempted by a doctor listening by ear to chest sounds, not exactly the most precise technique. Most people figured it out only once they were in the full symptomatic throes—and only once they'd lost their jobs and apartments due to fear they'd infect others. Edward Livingston Trudeau was the first American to promote isolation not only to protect the healthy but to help the sick. His Adirondack Cottage Sanatorium, encouraging lots of time in the fresh air, opened along Saranac Lake in the mountains of New York State in 1885.

X-RAYS ARE A NECESSARY TOOL

As long as there has been disease, there have been people who have cared for those who are suffering. We have records of dental disease—the diagnosis and the treatment of it—that are thousands of years old. Physicians even studied and tried to treat cancer before the Common Era. But for as many good intentions and successful treatments as there were, there were so many medical missteps that could have been prevented had X-ray technology existed. Patients suffered or died not only because of illnesses and injuries but also because of poor medical knowledge. Had more dental issues been able to be diagnosed sooner, "teeth" may not have been a leading cause of death for so many Londoners in the 1600s. Had patients with broken bones been able to step into an X-ray machine, they might not have had to suffer the agony of a bonesetter physically investigating their aching limbs and then manually pushing their bones back into place. Had President Garfield been X-rayed, the site of the bullet in his back would have been easily spotted and the bullet promptly removed. Instead, his death stretched over eighty days as doctors, fearful of where exactly the bullet was, unwittingly mistreated him. Cancers, respiratory diseases, and other ailments are more easily and successfully cured when they are diagnosed early. Physical exams can offer a degree of early detection in some cases, but X-rays show so much more, sooner. X-rays are not the answer to all health concerns, nor are they the right answer for everyone. But they do provide doctors, dentists, and other medical professionals with an invaluable second sight that saves lives.

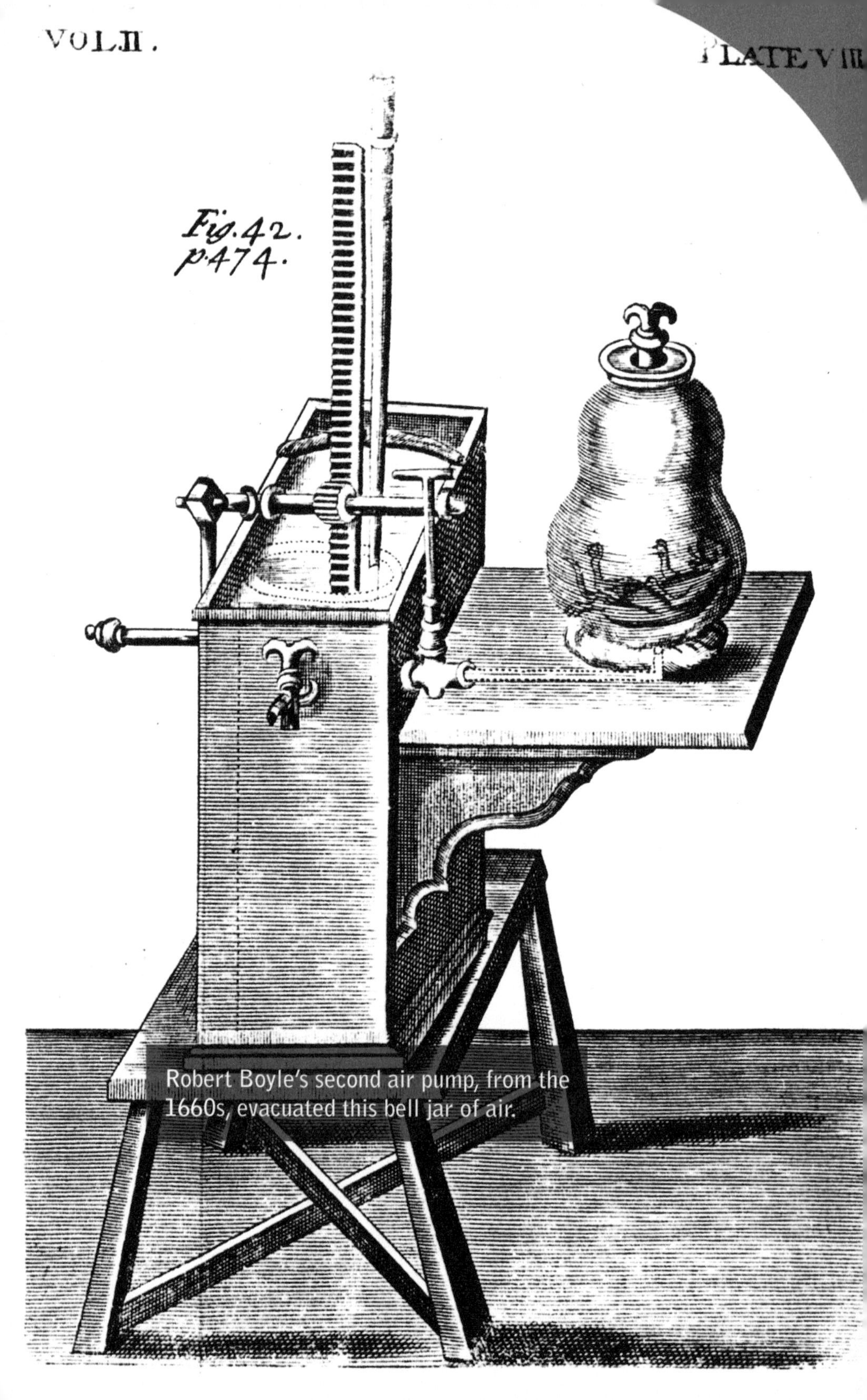

VOL.II.

PLATE VIII

Fig. 42.
p. 474.

Robert Boyle's second air pump, from the 1660s, evacuated this bell jar of air.

CHAPTER 2

The Science That Led to X-rays

The trail of discoveries and inventions leading researchers to X-rays, and then on to their widespread popularity, stretches from the 1650s. Quite surprisingly, it starts with the invention of the **vacuum pump**.

The VACUUM PUMP

The vacuum pump sucks air from a sealed space, leaving that space void of most or all air. We might think of vacuum-sealed food, food put in a special plastic bag that attaches to a vacuum that sucks all the air out of the bag so the bag wraps tightly around the food, thereby preserving it. Technically, what is happening is a decrease in air density within the space due to the suction. Then the absolute pressure of the remaining air drops, and the internal pressure level becomes lower than the outside. This is what creates a vacuum.

The First Pump

Otto von Guericke was mayor of Magdeburg, southwest of Berlin, Germany, when he developed the first vacuum pump in the 1650s. His pump consisted of a piston and an air gun cylinder with two-way flaps.

The movement of the piston agitated the gas and air molecules inside a container, creating an area of low pressure, which the molecules wanted to avoid. They did so by moving through the pump's outlet valve toward an area of higher pressure. When enough gas and air had been removed from the container, one simply closed the valve to prevent the molecules from being drawn back inward.

Magdeburg Hemispheres Experiment

To demonstrate the success of his device in 1657, von Guericke closed up two bronze hemispheres, which together formed a sphere, and suctioned the air out from within the sphere. (Von Guericke's early experiments used wooden casks instead of metal globes, but those imploded under the air pressure.) He then attached eight horses to the handle of one hemisphere and eight horses to the handle of the other, so that the teams were facing opposite directions, and ordered them to run. Sixteen horses could not separate two halves of a globe, each with a 1.2-foot (36.5 centimeter) diameter. Perhaps von Guericke took advantage of his high-ranking position in the town to put on this extravagant show. Since then, most Magdeburg hemispheres experiments are demonstrated with two strong humans trying to break the vacuum seal.

The Powerful Vacuum

As soon as scientists had a vacuum pump, they set about exploring all that it could do. The vacuum is a powerful thing, beyond just sealing shut a container such as the Magdeburg hemispheres. Von Guericke was the first to show that one cannot hear in a vacuum, that animals die in a vacuum, that a vacuumed environment can preserve fruit, and that a flame goes out when placed in a vacuum.

Otto von Guericke experimented to show that vacuums exist and that air has weight.

ELECTRICITY in a VACUUM

Electricity was one of the things scientists played with in airless states. As early as 1705, they found that electrical discharges could travel a longer distance in vacuums than outside them, but that was just the beginning. Research in this arena was another step closer to the realization of X-rays.

Michael Faraday

Michael Faraday (1791–1867) was one of the greats in the field of research. Though not all scientists of this time had the benefit of family wealth, most came from families with a certain amount of education, held steady employment with a fair wage, and enjoyed the acceptance in society that meant they had a safety net. Faraday, by contrast, grew up in distinct poverty. Faraday's mother had worked as a servant before marrying his chronically ill blacksmith father. Michael contributed to the family coffer by working as a delivery boy for a bookseller and then, after impressing his employer, apprenticing as a bookbinder.

Working with books all day every day, Faraday taught himself science. Through a series of twists and turns of fate, he met and impressed some great scientific minds of the time. One of the most famous of them all, Sir Humphrey Davy, invited him to be his assistant at the Royal Institution.

Faraday began to study electricity's movement in discharge tubes of gas. The wealthy were delighting in a toy version of this—as a snow globe puts a whole blizzard-bound city in a person's hand, these sealed lead-glass tubes displayed a personalized aurora borealis. Faraday attended a meal at the house of the vice president of the Royal Society that ended with such a show. He noticed a dark spot near one of the electrodes, the negatively charged **cathode**, which would be called the cathode dark space or the Faraday dark space. It would also come to be called Crookes space, after William Crookes.

The Contributions of Many Great Minds

Experimentation and innovation from lots of people were happening at similar times in relatively close geographic proximity. The history is full of tangled timelines as scientists, who sometimes worked as friends or colleagues in touch with each other by letter, conducted similar experiments. Sometimes they did so on purpose. Repetition of experiments by different scientists in different labs allows researchers to get up to speed in a way that reading a published report on the subject does not allow, double-checks the work, and may even add new findings to the experiment. Ultimately, Wilhelm Röntgen would publish his discovery of X-rays quickly, after only a few weeks of experiments on it, because he knew so many colleagues had to be close to the same finding he'd made. They all worked with the same materials on the same kinds of scientific phenomena.

William Crookes and the Crookes Tube

Armed with a degree from the Royal College of Chemistry in London; work experience at Radcliffe Observatory, Oxford, and the College of Science, Cheshire; and a bank account plump with an inheritance from his father, William Crookes in 1856 launched himself into full-time research at his own private lab.

One of his many inventions would come to be known as the **Crookes tube**, and Johann Hittorf would use it to discover cathode rays in 1869, though the term "cathode ray" was coined by Eugen Goldstein. The so-called tube is not an open-ended cylinder but a closed glass container. Picture a container that is shaped kind of like an elongated lightbulb or like a bottle with a short narrow neck, or perhaps like a tubular, not spherical, rubber balloon.

This glass tube is partially **evacuated**; using a vacuum pump, a scientist working with the Crookes tube empties it partially of air. Within this nearly airless state are two metal electrodes, conductors through which electricity enters or leaves an object or region. The negatively charged electrode is the cathode, and the positively charged one is the **anode**. Picture a simple double-A battery—those things on its ends act as electrodes. The one marked with a minus symbol is the cathode, and the one on the opposite side and marked with a plus sign is the anode.

When a high voltage of electricity is applied between the two electrodes, cathode rays project in straight lines from the cathode. These cathode rays are electron beams that light up the glass opposite the negative electrode in a brilliant glow. As he studied electrical discharges through the gas, Crookes noticed that even while the anode side of the tube lit up, a dark space grew around the cathode: the negatively charged electrode.

What was happening was that with little air and with voltage in the tube, even low-mass particles accelerated to high velocities. When they reached the anode end of the tube, in their excited state, they overshot the anode and struck the back wall of the tube. Those glass atoms were agitated, and their electrons moved at higher energy levels, causing them to fluoresce. Later researchers painted the inside back wall with fluorescent chemicals to make the glow more visible.

The first popular use of cathode-ray technology was for early television sets. Considering something familiar like a television can help to explain cathode rays. Within the set are airless glass tubes, vacuum pumped, containing a cathode in the form of a heated filament. From this cathode streams a ray, which we now know is made up of electrons that are negatively charged. Opposites attract, so the negative electrons want to get away from the negative cathode and move toward

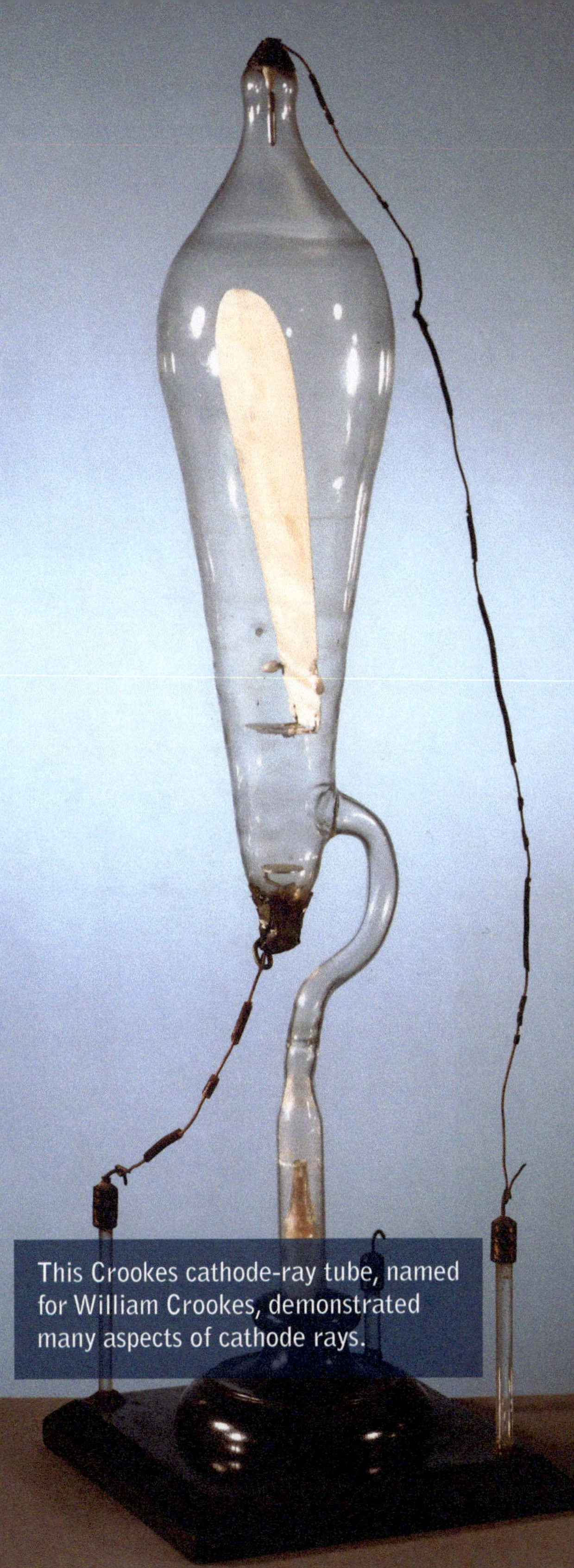

This Crookes cathode-ray tube, named for William Crookes, demonstrated many aspects of cathode rays.

The Explainable and the Not

Advancements came fast and furious in the late 1800s and early 1900s, when X-rays were discovered. People turned to both objective experiments and subjective faith for explanation. Even scientists seeking control within the seemingly whirling times might turn to belief in the otherworldly.

William Crookes, whose name would be on several discoveries and inventions related to X-rays, such as the Crookes tube, was born in 1832 to a tailor and real estate investor. Fifteen siblings would follow. Even in such a crowded house, William managed to focus enough to earn entrance into the Royal College of Chemistry at age sixteen. He hoped to study chemistry, but a chance meeting with physicist Michael Faraday changed his mind. He never obtained a graduate degree, yet his research did so much for the world.

Grief-stricken by the untimely death of his brother Philip in 1867, he started paying attention to **spiritualism**, the belief that people who have died can speak from beyond the grave to the living. He attended a séance in order to try to reach his brother and then spent several years researching the work of mediums.

There's great controversy about whether or not Crookes truly believed in a connection between this world and the spirit world. Some say he was prone to gullibility and had poor eyesight, both of which could have contributed to his believing to be true what he thought he saw. Some say he was in on a con with

Florence Cook in a trance at William Crookes's house, the "spirit" Katie King behind her, in the 1870s

medium Florence Cook in order to cover up his love affair with her (he was married with ten children). Some say he was purposefully duped—famous illusionist and stunt performer Harry Houdini even became involved, saying that medium Annie Eva Fay had confessed to him that she had tricked Crookes.

The scientific and the supernatural continue to make interesting bedfellows. In 2004, the powers of a teenager from Russia who claimed to have X-ray vision were put to the test. The Discovery Channel, producing a documentary on seventeen-year-old Natasha Demkina, asked experts to examine her story. Based on their preliminary research, they suspected she used not X-ray vision but "cold reading," a technique commonly used by psychics and other fortune-tellers. Had Crookes been alive today, how might he have worked in opposition to her—or with her?

the strategically placed positive of the anode—a path that runs the frantic electrons smack-dab against the television screen, which is coated with phosphor. Thus, the screen glows when struck by the beam.

TELEGRAPH

The era's new communication technology was key to how quickly and thoroughly X-rays advanced. The invention of the telegraph also involved some of the same scientific foundations and scientists as the discovery of X-rays.

Within days of Röntgen's announcement that he had scientific proof that X-rays existed, the news had spread around the world, and people were immediately starting to make use of the technology. That speed and thoroughness could not have happened before the telegraph, which Samuel Morse and others developed in the 1830s and 1840s. That heightened level of information dissemination meant people embraced X-ray technology, found new uses for it, and discovered its downsides sooner. The immediacy alone put it in a place of power within society, and the fact that it was developed so quickly by so many people in so many places strengthened it, setting it up to be the commonplace technology it is today.

Humans Have Never Not Talked

Long-distance communication was not a new idea when the telegraph was introduced. People in ancient China, Egypt, and Greece and the native peoples of what would come to be known as the Americas all used drumbeats and smoke signals to exchange information over hundreds of miles. But those methods, while ingenious and requiring no extra equipment, weren't perfect on, say, cloudy or windy days or at night. The later use of horses, carriages, and ships meant news could be

This illustration from around 1860 shows a woman using Morse code to telegraph a message.

delivered regardless of weather or time of day, but they took weeks to deliver their news.

The Telegraph–X-ray Overlap

Just as advancements in electricity and the understanding of magnets in part led to researchers finding X-rays, they also supported the invention of the telegraph. In 1800, the Italian physicist Alessandro Volta invented the battery. Before that, electricity could not be controlled—it couldn't be stored or used when and where we wanted. Twenty years later, Hans Christian Oersted, another physicist, in Denmark, demonstrated the connection between electricity and magnetism. Scientists saw the possibility of communication technology within a union of batteries and **electromagnetism**. Though Morse, along with his American colleagues Leonard Gale and Alfred Vail, is credited with the telegraph's invention, Sir William Cooke and Sir Charles Wheatstone in England developed their own unique-looking telegraph system, which ultimately was used for railroad signaling in Britain.

RÖNTGEN GETS READY

Certain scientific inquiries and inventions were necessary foundations for the discovery of X-rays. People had to be asking the questions and devising the instruments that, though meant for other purposes, would also allow them to see X-rays. But on the most basic of levels, in order for Röntgen to discover X-rays, there had to be a reason Röntgen was in his lab with the right equipment with which X-rays could appear. Wilhelm Röntgen was a fifty-year-old physics professor and recently elected rector at the University of Würzburg when his research took him in a direction it hadn't up until then, one of the many ways random chance led him into X-ray's path. Over the years, he'd published more than forty papers on everything from crystals to the effects of pressure on liquids, but he'd never looked at electrical charges in gases. He had been planning, for his next period of study, to expound on Hermann Ludwig Ferdinand von Helmholtz's predictions about the Maxwell theory of electromagnetic radiation. But Philipp Lenard's reports of his own research inspired Röntgen to move on that in a certain order, cathode rays first.

He purchased Hittorf-style cathode-ray tubes (these are pear shaped) and requested by letter Lenard's guidance in reproducing his experiments. Röntgen felt Lenard's work had brought the scientific community close to realizing Helmholtz's ideas. When he finished replicating Lenard's experiments, to similar results, he felt he was ready to take the science to a next logical step—proving what Helmholtz had suspected: the existence of invisible high-frequency electromagnetic radiations. He covered a thin sheet of paper with barium platinocyanide crystals that he hoped would help him to detect the presence.

Wilhelm Conrad Röntgen (1845–1923) won the first Nobel Prize for Physics in 1901.

CHAPTER 3

X-ray Pioneers

The science of, technology involved with, and applications for X-rays are diverse and ever growing. To narrow the field of pioneers and early innovators into one chapter is difficult. There are many more fascinating stories in addition to the stories of Wilhelm Conrad Röntgen, Émil Grubbé, Marie Curie, Ernest O. Wollan, and Martin Annis on the following pages. Yet these individuals are critical to our understanding of how X-rays changed the landscape of science and medicine.

WILHELM CONRAD RÖNTGEN: The FATHER of X-RAYS

Today we know that X-rays naturally occur in the world, so it may seem strange to think of a time when people didn't know that these rays of light existed. Before 1895, when Wilhelm Röntgen discovered them, doctors could rely only on external examination. Thanks to the scientist who once was expelled from school, health care has advanced a long way.

Education

Röntgen was born on March 27, 1845, in Germany, but he was raised in the Netherlands. His father was a cloth manufacturer and seller, and his mother came from one of the founding families of Amsterdam. They sent Wilhelm away to a boarding school, the Institute of Martinus Herman van Doorn. Though this did not delight Wilhelm—he wasn't much of a student—he did take solace in the school's beautiful location. The school is in the town of Apeldoorn, which borders what is now the Hoge Veluwe National Park, the Netherlands' largest privately owned conservation area. Though the park came into being in the early 1900s, after Röntgen lived there, the rolling hills of forests, grasses and flowers, and diverse animal population—from snakes to deer to the rare fritillary butterfly, the distant cousin of the monarch—were Wilhelm's escape from school.

As a teenager, Wilhelm started at Utrecht Technical School, but he didn't last there long, though not because of any fault of his. Another student drew a caricature of a teacher, and Wilhelm was blamed and punished for it. Fortunately, this didn't derail him. He went on to the University of Utrecht and the University of Zurich, earning his PhD in mechanical engineering. Though he hadn't done great in school, he had a sharp mind, and he did enjoy building mechanical objects. Along the way, he accepted the opportunity to study with some incredible physicists. So, though his life in education started a bit rough, Röntgen found his way to an impressive path.

Early Work and the Love of His Life

After so much diligent work, Röntgen was offered a series of professorships and chairs of physics departments at universities in Germany and France. In 1870, when he was only thirty-five years old, he started publishing findings of his research—at the time, on the temperature of gas, electrical characteristics of quartz, and the phenomena around oil beading on water, among others. He would always prefer to work alone, without even one assistant. This was not because he wasn't personable. Quite the contrary—he was known to be friendly, understanding, and empathetic. He just preferred to explore by himself.

His career gliding along, Röntgen was enjoying coffee in a café one day when his personal life changed forever. There he met Anna Bertha Ludwig, the daughter of the café manager. They married in 1872 in Apeldoorn, the location of Röntgen's first school. In 1887, they adopted their niece, the daughter of Anna's only brother. Josephine Bertha Ludwig was six when she came to live with the Röntgens.

X-ray Mania

With Röntgen's discovery of X-rays in 1895, he became a celebrity. Several cities named streets after him. Journalists coined the reaction "X-ray Mania." He won the Nobel Prize in Physics, one of the top awards in the world, in 1901. Despite being showered with praise, Röntgen continued as he always had: working, climbing in his beloved mountains, and entertaining friends with his family at their summer home in the foothills of the Bavarian Alps.

ÉMIL GRUBBÉ: EARLY RADIATION THERAPY SPECIALIST

When he announced his discovery of X-rays in 1895, Röntgen enjoyed a not necessarily common experience for scientists: his work was immediately accepted by the scientific community and celebrated by the public. Scientists try to poke holes in theories and question answers; they're not necessarily satisfied by the initial results of a colleague. And sometimes scientists are biased and find comfort in tradition and "how it's always

Pictured here in his Chicago office in 1959, Émil H. Grubbé pioneered X-ray treatments.

been done"; they're not necessarily willing to change their mind about what they've been taught or witnessed themselves. The public often glosses over scientific news in favor of gossip, politics, or sports. But people welcomed X-rays from the beginning. Émil Grubbé, a chemist and physician an ocean away from Röntgen at the time, in Chicago, was one of those people who not only opened his arms to the research but grabbed it and ran with it—almost literally.

Inventor

As Röntgen liked to devise and build his own equipment and other machines, so Grubbé was also a creator. In 1896, at age twenty-one, he was working as a chemist and assayer. He was always analyzing pharmaceuticals for work, and he also enjoyed experimenting hands-on with the burst of the electronic and mechanical gadgets of the era, on the exciting cusp of modernity. With this propensity for exploration, natural curiosity, and personal experience with Crookes tubes, induction coils, electric generators, and fluorescent and photographic chemicals and plates, Grubbé was the perfect recipient for the news of Röntgen's X-rays. Newspapers even printed detailed drawings of the apparatus Röntgen designed and built to make X-ray photographs. Grubbé was interested in how he could use this radiation caused by X-rays to cure cancer.

Childhood and Education

Born in Chicago on January 1, 1875, to parents who had emigrated from Germany, Émil Grubbé followed a fascinating educational path to his career. It is one that feels very nineteenth century in its extreme hardship but that also includes elements so many people today can still understand, including working multiple difficult jobs to earn money for family and school at the same time.

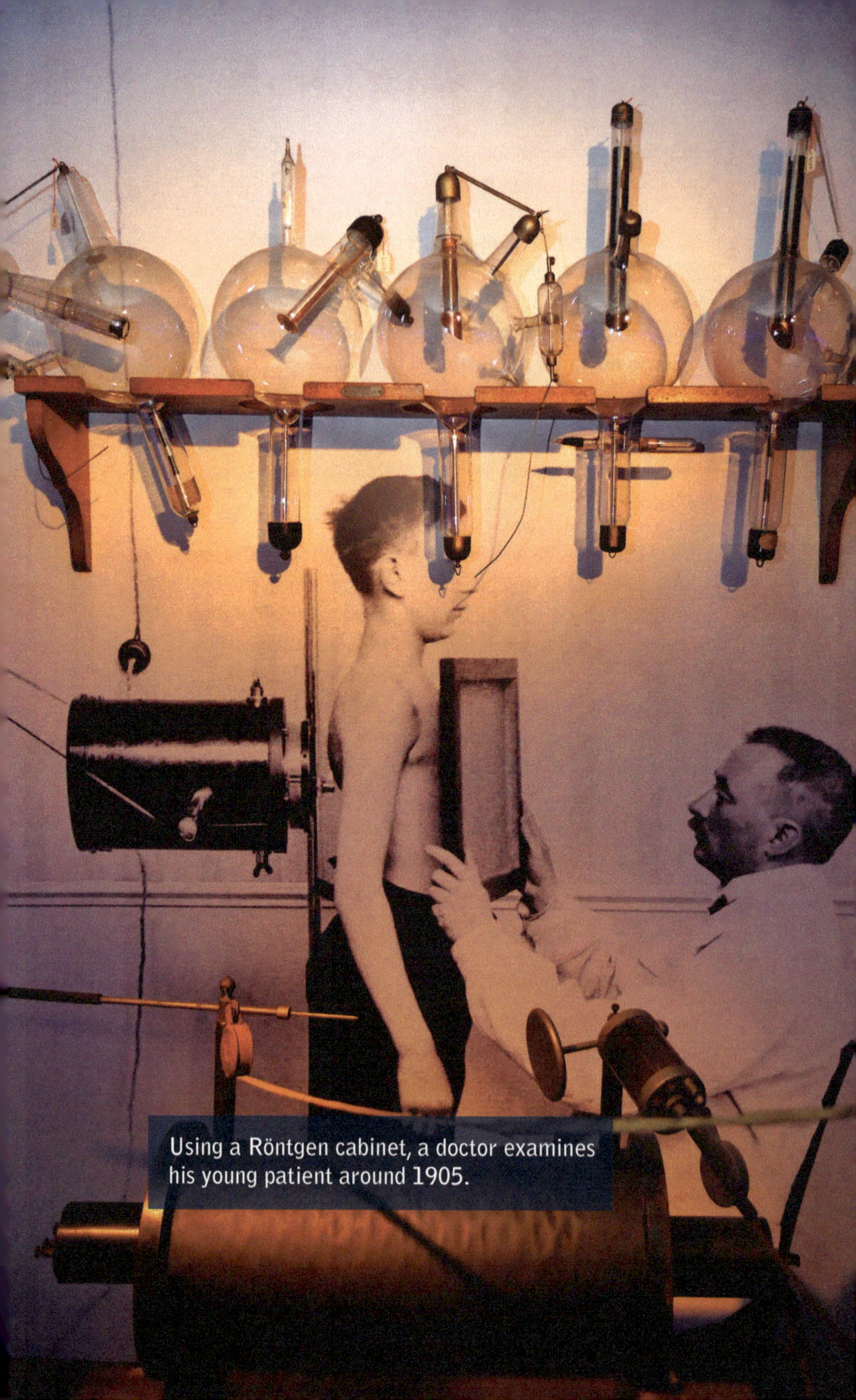

Using a Röntgen cabinet, a doctor examines his young patient around 1905.

Émil was a child before the concept of childhood existed in the United States. His first job was at age thirteen as a bottle washer and errand runner at a drugstore. Over the next year or so, he accepted a job with better working conditions and pay and, as it would turn out, the nudge he needed to start his career in medicine. He earned an impressive two dollars each ten-hour workweek helping in the office of the popular Marshall Field's department store in downtown Chicago. He did such a good job that Marshall Field himself took notice and got to know Émil. Noticing the assistant shared his love of science and medicine, Field encouraged Émil to go to school.

To attend medical school as a fifteen-year-old was not unusual in that era, but Émil hadn't really been attending any school up until the point, so he didn't meet the entrance requirements for the Chicago schools. He had to travel to Northern Indiana Normal School at Valparaiso to complete his studies. Today, Valparaiso is part of the Chicagoland metro; then, it would have been quite a trek. He went to school by day and worked as a watchman by night.

His grades earned him a place as an undergraduate in Hahnemann Medical College of Chicago in 1895, including an instructor position in physics and chemistry. His whole life, Grubbé worked a lot of jobs at once. After his initial steps with using X-rays to combat cancer, and finally graduating medical school in 1898, he held the professorial chair at Hahnemann in electrotherapeutics and radiography until 1919; taught at four different institutions in the city; maintained a private medical practice; and wrote, publishing about ninety research papers.

A Colorful Adulthood

When not working, he traveled the world and was particularly interested in volcanology. He achieved his goal of visiting almost all the volcanoes that existed then and witnessed the Mount Pelée eruption in Martinique, an island in the Caribbean, in 1902.

His personality was itself rather explosive. Curiously, he didn't claim public credit for his work with X-ray radiation and cancer until well after the fact and ended up spending the rest of his life fighting for the title of "father of radiation therapy," in addition to continuing to lead his full work and travel schedule. "He was colorful, flamboyant and occasionally saw himself in a somewhat better light than did his contemporaries," wrote the Chicago Radiological Society in its biography of him. Since 1970, Grubbé's estate has funded the Memorial Award of the Chicago Radiological Society, given in his honor and memory. In the will he left when he died of cancer in 1960, Grubbé set aside money to do this, administered by the University of Chicago.

Included in the instructions for the spending of this money was that someone should be paid to write his biography. The university's chairman, Paul Hodges, stepped forward "and discovered that the more he learned about Grubbé, the less he liked him." He said of the experience later, "if you're going to be fool enough to leave your money to have your biography written, then try to lead an exemplary life. Failing that, for God's sake, remember to tell your lawyer to stipulate that it be a positive biography." Ultimately in his biography of Grubbé, Hodges suggested, "Let us remember him rather for his self-sufficiency, self-confidence, tenacity, and a lust for life so great that it enabled him to contrive if not immortality at least long survival for the name Emil H. Grubbe."

MARIE CURIE: FIRST on the SCENE in WORLD WAR I

Marie Curie's Nobel Prize–winning work was inspired in part by Röntgen's X-ray discoveries, and after she secured her main project during World War I, she made use of X-rays directly to help fallen soldiers.

Marie Curie is pictured here three years after she became the first woman to win a Nobel Prize, for physics, with her husband, in 1903.

Petites Curies

As German troops marched toward their enemy in Paris during World War I, the French government relocated to Bordeaux. They asked Marie Curie to also transfer her precious, and dangerous, gram of radium—all that France had of the element—to the temporary capital as well. She did so but then decided that she herself had to return to Paris to help the French soldiers. She convinced the government to create a new position for her—director of the Red Cross Radiology Service—so that she could solicit money and supplies in order to set up France's first mobile military X-ray center. Having X-ray machines near the front lines would allow doctors to save more lives, as they would be able to quickly respond to injuries by seeing exactly where bullets, shrapnel, and bone breaks were in the enlisted men. By October 1914, the first of what would be twenty medical vans would be ready; they were repurposed and transformed cars from auto body shops and manufacturers. Soldiers would come to call these *petites Curies* (little Curies).

Curie's work relied on a foundation that Röntgen's discovery had helped to form, and she had lectured on X-rays at the Sorbonne—but Curie had never actually worked with X-rays! Since it was wartime, it was all hands on deck, and Curie knew that creating the petites Curies was not enough. She quickly taught herself how to drive an automobile and perform basic repairs; she also rushed through refresher courses on human anatomy and the use of X-ray equipment. With her seventeen-year-old daughter, Irène, at her side as her medical assistant, Curie set out for the front lines in late 1914.

Maria Sklodowska

Marie Curie is famously known by her married name, which she took after her marriage to Frenchman Pierre Curie, whom she met when she was a student at the Sorbonne in France, where he was a professor. Curie was Polish and born Maria Sklodowska on November 7, 1867. Her family nicknamed her Manya.

Fifty years before, Poland had come under the control of multiple other countries; Warsaw, where the Sklodowska family lived, for example, was governed by Russia, but another major Polish city, Krakow, was ruled by the Austrians. Still, Marie's parents raised her and her four siblings to be true to Poland. This nationalism cost them good jobs, but they accepted that and only urged their children to work harder. The Sklodowska parents were both teachers, so they saw power in learning. Marie graduated from high school at fifteen, with highest honors.

"Floating Universities" and "Secret Labs"

At the time, women were not allowed to attend the University of Warsaw, so Marie and her sister Bronya attended "floating universities," classes allowing women that met at night and always in different locations to escape interference by the Russian police. Clearly, the sisters realized, this was not a sustainable way to live. They had to go elsewhere to get a good education. They made a clever pact: Marie would work to help pay for Bronya's medical studies in Paris. As soon as Bronya had a job, she would return the favor. At a time when women were usually forced to rely on their fathers or husbands financially, this was a unique system.

In a nice twist of fate, Marie's temporary job would end up helping with her career. For three years, she was a tutor for the children of the owner of a beet-sugar factory outside Warsaw. In her spare time, she also tutored, for free, the children of the peasant workers and kept up on her own reading. She did all that at risk of punishment by the authorities, if they found out. The Russians didn't want the Polish people to become too educated so they banned a lot of different types of learning, including Polish-taught laboratory sciences. A chemist in the beet-sugar factory was as rebellious as Marie and gave her some chemistry lessons. This would be just the first "secret lab" she studied in—upon returning to Warsaw in 1889, she went multiple times a week to the "museum," an illegal chemistry lab for Polish scientists.

ERNEST O. WOLLAN: The FIRST HEALTH PHYSICIST

At the start of X-ray exploration, no one thought the waves could cause harm; they seemed to do so much good that people didn't even consider being cautious with them. Of course, as they learned about the benefits of X-rays, people came to realize the risks and potential dangers associated with them. This led to a related field, health physics.

Ernest O. Wollan was a cosmic ray physicist in academia when the Manhattan Project research team approached him, in 1942, to study the hazards of radiation. As part of this work, he developed the radiation symbol that is still used universally today.

He and his team developed appropriate monitoring instruments, physical controls, administrative procedures, monitoring radiation areas, personnel monitoring, and radioactive waste disposal. Their work formalized radiation safety and put into place methods so those protective measures could be enforced.

Born in Glenwood, Minnesota, in 1902, Wollan earned his PhD in 1929 at the University of Chicago by studying X-ray scattering. As a physics professor at North Dakota State College and Washington University, he researched cosmic rays in the United States, Europe, and South America.

It is interesting to note that Wollan was one of the few to participate in the start-up of the first nuclear reactor at Stagg Field considering he was also working on radiation safety at the time. This was a secret experiment on a squash court under the stands at the University of Chicago. Though the reaction couldn't even power a lightbulb, it was a key step in the development of the atomic bomb.

MARTIN ANNIS: INVENTOR of the FUTURE AIRPORT SCANNER

Scientists continue to find new applications for X-ray technology. In 2010, the US Transportation Security Administration (TSA) started installing whole-body scanners in airports. These X-ray machines show if a traveler is carrying something illegal, like a weapon, under his or her clothes. These machines use rays of light in one of two ways: bouncing X-rays off the body (backscatter) or sending out microwave-like energy. The end result is the same: a 3D image of the person in the machine, showing what's under the clothing.

Martin Annis invented backscatter—thirty years before the technology got its application. An MIT-trained cosmic ray physicist, Annis founded American Science and Engineering in 1958. In addition to developing projects for the space program, he was hired by the US Post Office and prisons to come up with ways to use X-rays to detect weapons, drugs, and other contraband. When he figured out a scanner he thought would help secure air travel, in the 1980s, the aviation industry turned him away, calling it an invasion of privacy that they couldn't get past the government or the public. Of course, today these types of scanners are in airports around the world.

Now that he's in his nineties, Annis is busy as ever. Sometimes he's amusing his six grandchildren; they really like the giant wooden airplane they can pilot, which he built and suspended with ropes and pulleys from the ceiling of his apartment. Sometimes he works on more X-ray inventions; in addition to improving the full-body scanner to be more effective and less costly, he's working on devices to detect breast cancer. He's always on his bicycle—even at his age; that's how he gets around his hometown of Cambridge, Massachusetts, and nearby Boston.

Marie Curie's Words

The following are thoughts from Curie on the power of science in times of duress as well as a short note from her to her daughter after Curie had arrived near a battle zone.

> The story of radiology in war offers a striking example of the unsuspected amplitude that the application of purely scientific discoveries can take under certain conditions.
>
> ...The great catastrophe which was let loose upon humanity, accumulating its victims in terrifying numbers, brought up by reaction the ardent desire to save everything that could be saved and to exploit every means of sparing and protecting human life.
>
> At once there appeared an effort to make the X ray yield its maximum of service.

Curie went on to describe how people came to accept her ideas and help in whatever way they could so that "the scientific discovery achieved the conquest of its natural field of action." For her, X-ray Mania—the term used to describe the sweeping enthusiasm for the technology, even among the mainstream public—wasn't about fads. It was about feeling a "more alive" "confidence" in research and about growing "our reverence and admiration for it."

Poperinghe, 24 January 1915
[Near Dunkirk]

Dear Irène,

After various wanderings, we've arrived here, but we can't make an attempt at working until we've made some modifications at the hospital. They want to build a shelter for the car and a partition to create the radiology room in a big ward. That all holds up the work, but it's difficult to do otherwise.

Curie goes on to give a rather blasé account of working in a war zone, mentioning the nearby fighting and the weather in nearly the same breath: "We hear the guns grumbling almost constantly. It's not raining, a bit of frost." She ends the note "With a hug" and signs it simply "me." To Curie, X-ray technology accomplishes what scientific advancement should: improving lives and encouraging further research.

Fluoroscopes such as this one were popular in shoe stores around the country in the 1950s.

CHAPTER 4

The Discovery Itself

No one was trying to find X-rays. They were an accidental discovery by a little-known researcher who named them with a non-name: he called them X, for "unknown" because he had no idea what these rays were or why he couldn't block them with glass, aluminum, copper, or even the walls of his lab.

The DISCOVERY of X-RAYS

When last we saw Wilhelm Röntgen, he was repeating his colleague Philipp Lenard's experiments and was moving into a new set of experiments, which would stem from what he found from those replications. He had brought into the lab a barium platinocyanide screen on which to test his results, though moving the screen close enough to the tube would be the last thing he did. He first set about controlling outside light, inside pressure, and any extra luminescence, establishing a workspace clear of things that might affect his results.

He covered the cathode-ray tube with black cardboard and closed the drapes, removing as best he could all competing light sources and blocking the luminescence that would be produced on the face of the tube. He was ready to begin his

This 1909 fluoroscopy scan happened before the dangers of radiation exposure were known.

experiment. He closed a switch on the tube, producing high voltage in an environment with the lowest possible pressure.

Had he been able to see under the black cardboard, he would have seen the illumination of the tube's anode end, under the cathode ray's effect. That wouldn't have surprised him. That was what cathode rays did. What they *didn't* do was what happened there in the lab, that evening of November 8, 1895.

A pulse of light out of the corner of Röntgen's eye caught his attention. He checked to make sure everything on this table was covered. The light remained.

He opened the switch, stopping the electrical flow, and the light disappeared. He closed it once more, and again there was the faintest hint of glow from across the room.

Röntgen struck a match, to investigate by its unobtrusive light. As he must have wondered, heart beginning to thump, the glow indeed seemed to be the screen of barium platinocyanide crystals. Which meant the glow had to be caused by something else in the room—the screen certainly wasn't glowing on its own—though the source seemed nearly as mysterious as if the screen itself had spontaneously lit up.

To Röntgen's knowledge at the time, barium platinocyanide didn't fluoresce without a little help from something else. Four years later, in 1899, organic chemist Friedrich Giesel would announce that it could act alone, naturally, no light striking it. This turned out to be true because barium ore has radium content. But how much radium it contains varies greatly by geography. Giesel's samples may have come from a region of what is now Germany and the Czech Republic that offers ore high in radium. In other words, it is not surprising that Röntgen didn't suspect the screen itself or even notice its self-induced fluorescence earlier—the intensified glow under the effects of the X-ray was barely visible.

Except for the table of cathode-ray tubes, the room wasn't in use. That could be the only source of the disturbance. But the tubes were shielded with black cardboard. Even if their cathode rays hadn't been barricaded, Röntgen knew they shouldn't be able to travel all the way to the screen about 3 feet (1 meter) away.

Especially for such a reserved person as Röntgen, what he did next was downright wild. He added a sheet of cardboard between the tube and the screen, and then another. Then he set a thousand-page book between the tube and the screen. He slid a wooden board from a shelving unit in there, too, for good measure. The screen continued to shimmer. Dinner with his wife was forgotten.

He moved the screen itself closer to the tube, which intensified the wisp of light into a pulsing mass. We know from others' experiments that it was green in color; Röntgen didn't note color in his log because he was color-blind. We also know from later research that this intensity happened because of the way early cathode-ray tubes were designed. They allowed X-ray intensities to fluctuate. Röntgen moved the screen back, to a distance of 6 feet (2 m); still the glow, though fainter, remained. He kept at this until his stomach reminded him to eat.

Cautious Optimism

Röntgen would write in his lab notes that he wasn't sure what he'd seen, if anything. This was understandable—he'd seen the glow only once and not even while looking directly at it. Röntgen's discovery was caused by layers of accident. The glow of the barium platinocyanide screen that indicated something strange was not strong; it was the faintest of glimmers. Another convenient accident was that the X-rays hit the screen at a moment Röntgen was not facing the screen—oddly, the perfect direction from which to see the fluorescence. He was looking where he thought he should have been looking, at the experiment he was purposefully conducting. He caught the glimmer of the screen indirectly and through the sensitive part of his retina, the parafoveal region.

Röntgen detected X-rays on November 8, 1895.

Looking where one was supposed to look was one of the reasons other scientists working in the same field with the same equipment—even the same experiments—didn't learn about X-rays before Röntgen did. So many brilliant minds had been close to the rays; they'd just all managed to explain away what they saw. Otherwise, they were so focused on their work, they were not even fully able to register what they saw. In the 1880s, Crookes developed what would later be recognized as a prototype for the modern X-ray tube. However, during his research with it, he thought something was wrong, not right. He noticed that unexposed photographic plates would sometimes appear fogged, but instead of asking himself why, he returned the plates as defective to the manufacturer. Twice Lenard's experiments brought him close to discovering X-rays, but Röntgen happened to be the one who channeled his intelligence; keen observation (which was perhaps due in part to his color-blindness, so his sense of sight honed itself in other directions); and open-mindedness into discovery.

Further Experimentation

That dark night of November 8, Röntgen barely ate or spoke with his wife, Anna Bertha, and he returned to the lab in short order after eating and saying little. He still had no idea what was happening, but he knew something was. He replaced the paper and wood barriers with thin plates of aluminum and copper—no change was evident—and then lead and platinum. Those two blocked whatever was causing the screen to fluoresce, so Röntgen focused on experimenting with lead.

He set a sheet of it in front of half of the screen, and while that half did not glow, the unblocked half did. The way it looked, the way the screen became so crisply either light or dark, with no bleeding of either lightness or darkness at the edge of the lead plate, made Röntgen think of the effects of rays of light.

It was time, thankfully, for yet another stroke of accidental fortune. Still testing all manner of lead shapes, Röntgen held a disk up between the tube and the screen and saw not only a round patch on the screen, where the disk blocked the rays, but also his own hand. More exactly, he saw the bones of his hand! Once he regathered his wits—imagine having seen your own ghost—he extended his empty hand forward again. And there, again, was his spectral hand projected on the screen, five bony fingers dancing.

Röntgen promptly paused his university teaching and administrative duties to spend the next six weeks in his lab. Röntgen may have even suspected he was uncovering a game changer of global proportions; if he did, he was right.

He set about following the testing method his science education had trained him in, asking questions and following experimentation through to answers. Did these new rays move in straight lines, like cathode rays did? Yes, he determined, tracing the path of the fluorescence from its source. Were they refracted? He could not refract them with water or carbon bisulphide in mica prisms. Ebonite and aluminum prisms refracted the rays on photographic plates but not against fluorescent screens. Could he concentrate the power of the rays? Not with ebonite or glass lenses. Since these rays seemed more able to pass through substances than be stopped by them, what were some of the most extreme possibilities? He found that X-rays could move easily through thick layers of powdered rock salt, electrolytic salt powder, and zinc dust, and while those don't sound as sturdy as metal, remember that visible light can't pass through grime. That's because visible light is subject to reflection and refraction. Röntgen deduced that X-rays are not.

He found that he could control the point from which X-rays originated by controlling the cathode rays. X-rays stem from the cathode's strike point against the glass tube. While X-rays are unaffected by magnetic force, cathode rays are powerless against it. So, he could use magnetism to move the cathodes and, therefore, move the X-rays. This also helped demonstrate that there were two different types of rays present.

Röntgen finally breathed word of what he was doing to another person on December 22, 1895, when he led his wife into his lab. They emerged later, Anna Bertha exclaiming, "I have seen my own death!" Her husband carried spooky proof of her words: an image of the bones of her hand, the ring she wore clear upon it. The time it took to produce a radiograph in those early days of X-raying was much longer than it is now. Anna Bertha had to hold her hand still under the beam's glare for fifteen minutes. It took a quarter of an hour to watch her bones appear.

BUT WHAT ARE X-RAYS?

In the late 1800s, electricians and physicians alike pondered what X-rays were. Albert Michelson, who would become the first American physicist to win a Nobel Prize, thought they were vortices in the ether. Thomas Edison and Oliver Lodge, who played a big role in radio technology, leaned toward acoustical waves.

Now we know that X-rays are light. Specifically, they're energetic photons emitted by electrons. X-rays are a form of electromagnetic (EM) radiation, like radio waves, microwaves, infrared light, visible light, ultraviolet light, and gamma rays. X-rays vibrate at much shorter wavelengths, shorter than all of the above but gamma rays. They also vibrate at higher frequencies, allowing them to pass through all sorts of solid matter, including human tissue, clothing, and packing materials. The shortest wavelengths of visible light are around four hundred nanometers, but X-rays are one nanometer and smaller. X-rays also deliver about ten million times more energy than a radio wave does, which is at the least dangerous end of the wavelength spectrum. This is why X-rays are a powerful form of energy—in ways both bad and good.

THE ELECTROMAGNETIC SPECTRUM

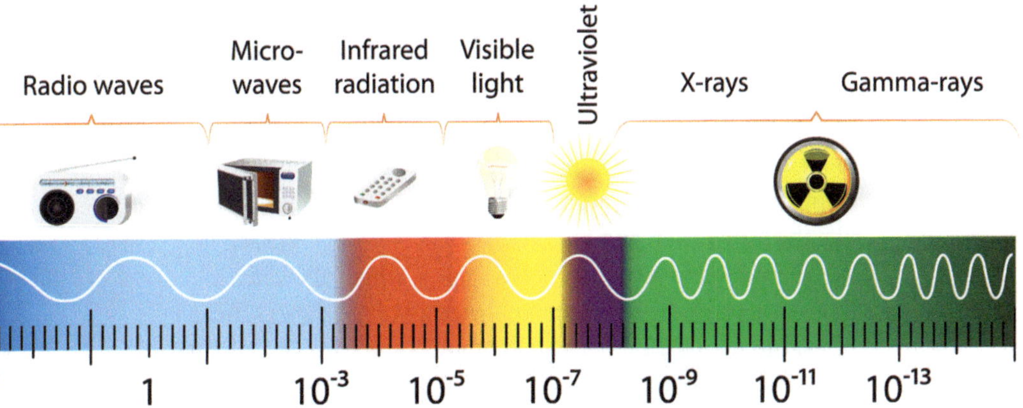

The electromagnetic spectrum vector diagram orders different types of electromagnetic radiation by wavelength.

Consider how rays of sunlight cannot go through you. You cast a shadow within the pool of light that continues on around you. That's how X-rays show images of bones so well—the waves pass through a body's soft tissue and are blocked by bones, leaving images of them as white shadows on the film.

Subcategories of X-rays are mammograms, dental X-rays, contrast X-rays, fluoroscopy, and CT scans. (MRIs and ultrasounds do not use X-rays, though.) X-rays are well suited to taking pictures of dense items, like bones. The X-rays used for this are hard X-rays, which carry more energy than soft X-rays and therefore aren't easily absorbed when traveling through air or water. Large-particle accelerators produce the brightest X-rays. Mammograms and dental X-rays focus on certain parts of the body, the breasts and the mouth, respectively. CT scans are well suited to photographing organs like the heart, liver, and kidneys.

X-rays led to X-ray **crystallography**, which allowed a clearer picture of the molecular structure of substances, from viruses to medications as well as DNA. X-rays are also leading to a new generation of lasers that are so bright and burst for such a brief span of time that they're perfect for creating stop-motion movies of natural phenomenon that happen at an incredibly small scale, like the formation and breakage by atoms of molecular bonds. We can see not just the still image of a chemical reaction at different points but the reaction as it happens, at its molecular level. This provides us with a full understanding of what's happening.

LIKE WILDFIRE

As sudden as its discovery was the spread of the news about the X-ray and its popularity with scientists, doctors, and the public alike. Part of that haste was Röntgen's doing. He himself couldn't believe he was the only scientist aware of X-rays. Röntgen was working at a time when so many scientists were focused on rays of light. He spun through all sorts of paranoid thoughts, including that he'd made not a scientific discovery but a spiritualistic one. As big as science was at this time, so was spiritualism—even scientists such as Crookes had come to (at least claimed to) believe that the dead could speak to the living. Three of Röntgen's closest friends had died recently, so maybe he was more attuned to the spirit world, he thought. Röntgen pushed that idea aside for a more reasonable one: that he really was sitting on a discovery that was soon to be made by any one of a number of colleagues of his. He needed to move if he was going to see any professional benefit from all the work he'd put in.

On December 28, 1895, Röntgen revealed what he'd learned to the Würzburg Physical Medical Society. He convinced them to publish his handwritten manuscript on what he'd learned. By January 4, 1896, word had reached the Berlin Physical Society. The local newspaper, the *Wiener Press*, picked the story up on January 5, and by the next day, the news was flying by telegraph around the world. On January 16, the *New York Times* exclaimed over how this "photography" would transform surgery and expressed what so many were thinking: when would the details of Röntgen's work be available in English?

Within weeks, X-ray Mania was sweeping the world, mostly in terms of immediate medical advancements but also in silly exploits. The famous Lord William Thomson Kelvin, creator of the Kelvin temperature scale, raised the only fuss. But his doubt about the discovery's legitimacy was reasonable—not two months before, these unknown rays hadn't even been dreamed of. Even Röntgen couldn't disagree that things were rushed. He expressed to Anna Bertha his worry that he'd been too hasty in his announcement and that he had not experimented enough. As he mailed copies of the "shadowgraph" of her hand to prominent scientists across Europe, he worried that fault would be found with his work, destroying his career.

That wouldn't happen.

The Maastricht Machine

Röntgen's announcement sparked immediate copycat experiments, though Röntgen likely didn't see it that way, as he believed the technology belonged to everyone. Still, it was amusing that people all over the world were jumping into X-ray Mania with such enthusiasm, not just with novelty items but with working X-ray devices.

In 1896, H. J. Hoffmans and L. Th. van Kleef made an X-ray machine in Maastricht, the Netherlands, using parts that Hoffmans had lying around the high school where he was director. He had been trained as a physicist, and his friend van Kleef was director of a local hospital. Copying Röntgen some more, they also took images of a female hand, van Kleef's twenty-one-year-old daughter's.

The machine ended up in hospital storage until 2010 when one of the doctors of the Maastricht University Medical Center dusted it off for a television program on regional health care. His colleagues at the hospital were intrigued, and the next year physicists, engineers, and radiologists got it working again. They upgraded the battery and some wires, but they kept the rest of the original materials, including a wire-wrapped iron cylinder to transfer electricity and a glass Crookes tube with metal electrodes at each end.

After an hour of fiddling, the machine glowed. Just as the discovery of X-rays more than one hundred years

This 1890s X-ray installation is one of the oldest in the world.

before had been accidental, so was the success of these researchers with this first-generation X-ray machine. They didn't know what was wrong with the old machine or if it could still work. They just played around with it and were lucky they hit on the solution before they decided the machine was beyond repair.

Bringing this X-ray machine back to life wasn't just for fun. Original X-ray systems are lost to history, so without testing a machine from then, there would be no way to know how powerful those turn-of-the-twentieth-century machines were. This group of modern-day researchers found that the Dutch machine gave the skin a dose of radiation fifteen hundred times greater than the same image would require today, and while today's exposures take twenty-one milliseconds, this machine took ninety minutes.

Medical Uses

In 1896, a professor at the University of Pennsylvania noted to the *New York Times* how he could see using X-rays to diagnosis everything from kidney stones to cirrhotic livers—really, "the surgical imagination can pleasurably lose itself in devising endless applications." Two weeks after Röntgen's announcement, Friedrich Otto Walkoff produced the first dental radiograph. He wrapped an ordinary photographic plate in rubber, placed it between his teeth and tongue, and lay on the floor for what he described as twenty-five torturous minutes. The next year, he and Fritz Giesel established the world's first dental röentgenological laboratory, a lab dedicated to taking dental X-rays.

On January 19, 1896, only three weeks after Röntgen's announcement, the first clinical X-ray was made in the United States. Eddie McCarthy, a boy who fell while ice skating on the Connecticut River came to have his wrist X-rayed and became a ground-breaking patient this side of the Atlantic.

A member of the nearby Dartmouth Scientific Society, photographer Howard Langill, read an article about all the ways European medical professionals were making use of X-rays. He inspired Frank Austin, an assistant in the Dartmouth College physics laboratory, to make use of the dozen Crookes tubes at the college, considered one of the strongest collections in the country at the time. Langill provided twenty-five gelatin plates. Only one of the tubes produced X-rays. It differed from the others because of the mica coated with phosphorescent material set inside it.

Edwin Frost was a professor of physics and astronomy at Dartmouth College, so he knew about Austin's experiments—and he was also the brother of fourteen-year-old Eddie McCarthy's doctor. So, a couple of weeks after McCarthy's fall, Dr. Gilman Frost brought his young patient to his professor brother's lab. The latter Frost examined McCarthy's wrist and took an X-ray. After exposing the gelatin plate for twenty minutes, they could see the fracture clearly in the X-ray image.

By 1896, one of the first radiology departments in the world was created, at Glasgow Royal Infirmary, and there John Macintyre produced the first X-ray of a kidney stone, a frog's legs kicking, and a strange object swallowed by curious child (a penny).

EVERYONE'S TECHNOLOGY

Everyone around Röntgen urged him to patent X-rays and therefore reap monetary rewards from their use for years to come even well after his death. His general work philosophy—take pride in your work but don't allow egoism and arrogance to lead to snobbery in your research—wouldn't allow him to do this. Specific to the discovery of X-rays, though, he believed that no one individual or group could—or should—own such a world-changing scientific advancement. To prove that point even further, he donated the check he received as part of his Nobel Prize to the University of Würzburg—$1.2 million at today's rate of exchange. The world, however, continued to give him claim to the invention, in name at least. Though X-rays are referred to most commonly by their mystery-lovers name, they are also known (most often in German-speaking communities) as Röntgen's rays, and the images they produce are called röntgenograms.

ILL EFFECTS

Almost as soon as medical practitioners started using X-rays, they started noting their ill effects on human health. Röntgen didn't, likely because he kept the X-ray beams partitioned in his lab with a zinc box and lead plate. He intended to protect his photographic plates from being exposed and ended up protecting himself from scattered rays as well. Also, the very first machines, in the fast rush of first-generation X-ray machines, weren't as dangerous as the ones soon to come.

In 1886, the owner as well as an employee of an X-ray tube manufacturer reported itching and burning in their hands. A new doctor, HD Hawks, started a demonstration of a powerful X-ray machine. After four days, the skin on one of his hands became very dry; then his hand, looking like it had been severely burned, started to swell. After two weeks, the skin sloughed off, his knuckles ached, and his fingernails stopped growing. Hawks covered his increasingly ailing hands in petroleum jelly, then gloves, and then foil. His hair started to fall out, and his chest appeared burned as well. His bloodshot eyes blurred. And William Levy, who wanted a ten-year-old bullet finally removed from his skull, sat within a triangulation of X-ray exposure focused around the right side of his head for fourteen hours so that imaging could locate the bullet. Within a day after he left the lab, his entire head had blistered; after a few days, his lips and right ear were severely swollen and his hair had fallen out on the right side of his head. Even in the early days of X-ray usage, there were many reported cases of skin that looked burned and blistered and, at the extreme, limbs had to be amputated due to this damage. We know now that these kinds of injuries mean people were being exposed to at least 1500–3000 rads, a unit that measures radiation absorption.

But the biggest problem, in a way, was that while people were reporting these incidents, they weren't ceasing their unprotected prolonged exposure to X-rays. People noticed concerning qualities in their health, but they didn't want to believe they were symptoms of anything serious linked to X-rays. They also didn't know that X-rays produce a cumulative effect—even minor doses, if administered regularly over time, can equal a big dose to the human body. The tube manufacturers said they'd keep a close eye on what happened next. After six weeks, Hawks showed signs of improving health and he joked about what he'd been through. Even the swollen, bleeding, half-bald gunshot victim, Levy, said that he'd been satisfied with the procedure for showing exactly where the bullet was in his head.

The many medical professionals who suffered on through pain, disfiguration, and childless marriages were called "martyrs to science through Roentgen rays," in one 1936 collection of essays. A doctor who died in 1960 wrote in his autobiography the he would "die a victim of natural science, a martyr to the X-rays."

It took death for people to start questioning the risks of X-rays. Thomas Edison's assistant, Clarence Dally, logged a lot of hours working on Edison's X-ray lightbulb. After years, his hair fell out and the lesions now common to his skin stopped healing. Burns on his hands became cancerous. His left hand was useless, it was so swollen and painful, but instead of ceasing work, he just relied more heavily on his right hand. After six years, the pain in his left hand was so great that he had to have it amputated. Eventually, both his arms were amputated, and then he still ended up dying in short order, in October 1904, at only thirty-nine years old. He was the first to have died as the result of X-ray exposure. In 1927, after thirty years of X-raying jaws and heads, dental radiologist Giesel died of metastatic carcinoma (a type of aggressive cancer) from radiation exposure to his hands.

The POWER to CURE

Those pioneers of X-ray technology took many unnecessary risks; as soon as they understood that bad things happened to the human body quickly and regularly when exposed to X-rays, they could have taken precautions while still advancing the science. Though they probably didn't have to go to the extremes that many did, it was useful that they learned how dangerous X-rays were. These killer rays are also healing rays: scientists quickly realized that if they could destroy healthy cells in people, they could do the same to unhealthy ones, perhaps killing the disease those people faced. Cancer had previously been detected only at later stages in the disease's development in a body and "cured" only by removing the infected body part, if that were possible. Suddenly, with X-rays, cancer was able to be seen earlier and removed in an ultimately less damaging way.

X-rays created the possibility of early detection of TB as well. The rod-shaped bacterium that causes the disease, *M. tuberculosis*, casts shadows on and causes spots in the lungs. More solid organs like the heart appear as whiter areas on X-ray film, like bones do. These organs offer greater resistance to rays and bounce them back. But lungs are full of low-density air, so they show up as dark gray to black, perfect backdrops for noticing even the hint of lung disease. With more blood rushing to the lungs, because TB has an inflammatory component, even the first traces of the disease can be seen in the X-ray image.

PUBLIC USE

Frank Austin, the lab assistant at Dartmouth, made a portable X-ray machine to use when he was demonstrating the technology's usefulness to doctors around New England. Of course, it was too cool not to bring out at parties—his daughter remembered that he brought it out during at least one of her birthday parties to X-ray the children's hands, causing "loud shrieks of excitement." What inspired him to do it was that they were all wearing fancy costume rings as favors—it was like a whole party of little Anna Bertha Röntgens. They were not alone in their joyful enthusiasm for X-rays. Carnivals and theatrical shows started to use X-ray machines for their entertainment value. The 1904 World's Fair in St. Louis, Missouri, exhibited an X-ray machine.

X-rays were also used in the consumer realm. As explained in an article on X-rays in advertising in the journal *RadioGraphics*, companies have always sold products using

A World War II poster shows a world-traveling sailor who makes time for routine X-ray appointments.

whatever term currently screams "advanced" and "edgy." That might be "laser," "atomic," or "nuclear"—in the late 1800s and early 1900s, the word was "X-ray." People knew X-rays could see through skin and had heard they were going to cure cancer; they thought that in the distant future we'd use X-rays to spot broken hearts and raise the dead. So why couldn't they sell headache medicine, golf balls, stove polish and furniture polish, antiseptic ointment, batteries, razor blades, coffee grinders, lemon juicers, whiskey, and a vacuum drinking fountain for chickens using the word "X-ray"? Advertisers didn't necessarily claim X-rays made or fortified their products—they basically branded them with the word. "X-Ray Raisin Seeder, The One that Seeds" boasts the surviving letterhead for one product. The word implied the height of modernity, unseen strength, and the ability to see through to the truth.

In the 1920s silent film *General Personal Hygiene*, produced with approval by the US surgeon general, X-rays show how to fit feet properly with shoes. Shoe store fluoroscopes began appearing in stores to help every clerk fit every person properly. These were wooden boxes about 4.5 feet (137 centimeters) tall. An X-ray tube and fluorescent screen were placed inside, and the box was lined with .07-inch-thick (2-millimeter) lead sheets. A fluoroscopy is a continuous X-ray image, offering real-time monitoring—it is not a static picture of a moment in time. The customer, usually a child, particularly entranced by the idea of seeing his or her skeleton, would try on a pair of shoes and then slide his or her feet into the two holes at the base of the cabinet. There were viewfinders enough for the clerk, the mother, and the child to see how the shoes fit the wiggling green-yellow glowing toes, though often kids tried the machine out by themselves and not for the purpose of actually buying shoes—they were just enjoying some free entertainment after school. In the early 1950s, there were ten thousand of these X-ray machines in stores around the United States.

As the popularity of fluoroscopes reached its height, scientists started expressing concerns about the health risks they presented. In a 1948 evaluation of two hundred fluoroscopes in the Detroit metro area, forty-three were found capable of exposing shoppers and employees to excessive radiation—by standards of the era. The problem would be considered even worse today. Doses ranged from sixteen to seventy-five roentgens (a unit of exposure to X-rays) per minute, and people were told to stand there for two minutes. Radiation was measured as far away from the wooden cabinet as 10 feet (3.04 m), so even clerks and customers not looking at their feet were exposed. Two *New England Journal of Medicine* articles in 1949 described potential issues with foot development in children who used the machines, and an article in a 1953 issue of *Pediatrics* advised against young people standing in them. Between January 1957 and 1970, thirty-three states banned the devices; the last one in the country was unplugged in 1981.

The Hubble Space Telescope, Spitzer Space Telescope, and Chandra X-ray Observatory use infrared and X-ray light to reveal the Milky Way's core.

CHAPTER 5

The Influence of X-rays Today

Many hidden medical conditions of centuries and millennia past weren't diagnosed in time to save patients. However, scientists today are using X-rays to understand more medical conditions; their advances help patients of the present—and the future.

USING MODERN SCIENCE to UNDERSTAND the PAST

New Findings in Ancient Dentistry

Without X-rays, our most exact knowledge about ancient dentistry comes from the written record. Researchers can't necessarily see with the naked eye what caused the jaw pain of someone who lived and died long ago, so they rely on the notes of the physician or dentist who treated that person. That remains incredibly helpful to medical research; some things, such as conditions of the soft tissue, can't be seen even by X-rays because those tissues have long ago decayed. Written records and X-rayed bones, together, help researchers examine skulls found at archaeological sites more thoroughly than ever before.

They've determined that the earliest dentists lived in the areas of China, Egypt, India, modern Italy, and Japan. In 2012, researchers discovered the oldest dental filling yet: 6,500 years ago, one of the world's first dentists fixed a cracked tooth with beeswax. The skull of the patient was found in modern-day Slovenia. Much more recently, but still a very long time ago, a dentist 2,100 years ago cleverly soothed a patient with numerous cavities by packing the deepest one with woven plant fibers. The lucky recipient of that top dental technology of his time lived in Egypt.

Cancer from a Million Years Ago

Using **micro-CT** imaging, researchers diagnosed cancer in a toe bone found in South Africa and dating to 1.6 to 1.8 million years in the past. This technology allowed scientists to note differences in density within one small bone and review the bone from different angles. The telltale growth pattern of the bone tissue in a cauliflower design led the team to identify the person's disease as osteosarcoma, a cancer still afflicting humans, primarily children and young adults.

Understanding the rates and types of cancers from centuries and millennia past is not merely an academic exercise. This knowledge helps us think about prevention and causation of cancers of today. The same archaeological team that found the South Africa toe also unearthed a vertebra that paleonthologists (scientists who use fossils to learn about ancient life) determined had been marred by a tumor. At 1.98 million years old, dated by its host bone, that tumor is the oldest known benign tumor. It is proof not only that human genes have long had the capacity for cancer but that there have also been biological control mechanisms that cause tumors to self-limit in strength or size. Therefore malignant tumors, those bearing cancer, as well as benign ones have basically always existed. Because we're able to study the remains of past cancer sufferers, we can better understand what causes different cancers. For example, from historical and archaeological study,

A watercolor showing colloid cancer on the toes

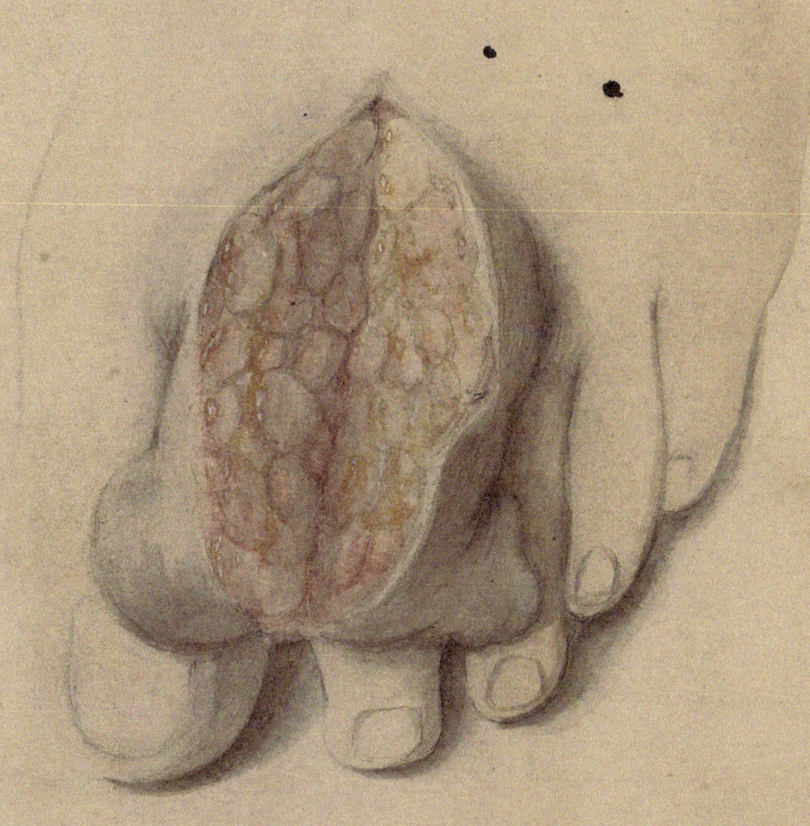

we know that stomach cancer was more prevalent until the late 1800s, likely because of carcinogens found in food preservatives at the time. Thinking about what came before helps us understand what we face today and may face tomorrow.

REVEALING ANCIENT LIVES

X-rays are useful not only for determining what diseases people of ancient times died from but also how their bodies were treated after death. In 2015, it was discovered that a Chinese statue of the Buddha contained an entire mummified monk, perfectly fitting the statue, as though he were in a mold. The CT scan showed not only the monk, who likely died around 1100 CE, but the cavities where his organs used to be, now filled with scrolls of text. And as easily as we can see through skin to our bones and organs with X-ray technology, we can see through the hundreds of yards of linen used to wrap ancient Egyptians after they died. Thanks to X-rays, archaeologists and other researchers don't have to destroy the mummy's outer wrapping to see inside.

That outer wrapping, cartonnage, is itself important to understanding past societies and individual lives. Cartonnage is layered papyrus, plastered smooth with gesso so that the surface could be painted and molded into masks and other accessories for the mummy or even into the interior coffin that was in fashion for a time for burials of elites. Papyrus was an early kind of paper made from the pressed pith of the papyrus plant and used exactly as we use paper today: something on which to write things. Undertakers for the middle class of ancient Egypt used recycled papyrus the way we reuse paper, rarely examining what's written on it—just adding pieces of it indiscriminately to our projects. This means that there are a whole bunch of stories from other people's lives, not just that of the mummified person, to be had by examining those recycled scraps of paper decorating the mummy. Everything from hate mail and rambling diary entries to business contracts and purchase receipts to excerpts of drafts from literary writing has been found. We have so little written record of the everyday lives of common people from even the near pre-internet past, let alone from the distant past of ancient Egypt. What's especially exciting is that many of those scraps of papyrus tell us about women's lives, which have been left out of most histories.

Unfortunately, until recently, we had no good way of separating these layers of paper while preserving the text written on each sheet. One technique has even been to carefully massage the cartonnage in soapy water, trying to separate the pages from each other with the gunk-loosening power of suds. That is clearly an invasive and incredibly risky way, and scientists no longer practice it and have been looking for something better.

In 2016, a private imaging expert met with archaeologists, physicists, and engineers from the University of California, Berkeley; Duke University; Stanford University; and University College London at Lawrence Berkeley Lab in the San Francisco Bay Area. They pooled the knowledge from their diverse fields to consider the most effective way of examining ancient papyri.

In addition to the low-tech soap solution, there are lots of high-tech possibilities for investigation. The private imaging expert present at this informal summit has had success with multispectral imaging of artifacts. He aims light—moving in increasing wavelengths, from purple to red to infrared—at artifacts. Each type of light bounces off each type of material composing the artifact in a unique way; with each distinct bounce, researchers can see a distinct facet of the artifact, some of which aren't visible to the naked human eye. Put those images together, and there's a more complete picture of what composes the artifact or what is hidden within it. For example, under multispectral imaging, "invisible" marks like fingerprints on paper can also be seen. Also, ancient Greek mathematician Archimedes's formulas were distinguished from the smears and smudges later dirtying the piece of paper on which he'd written. The problem with multispectral imaging of items such as mummy masks is that masks are not made of one whole sheet of papyrus—they're strips and pieces of different documents layered with the text going in all different directions. Over time, creases and wrinkles may have twisted the lines of text further.

When a physicist studying the photosynthesis of the spinach plant read about the Archimedes discovery, he had an idea for a way of studying ancient text on complicated surfaces such as cartonnage. His research relies on the detection of trace metals—and so might the "seeing" of obscured text. If the ink includes bromine or iron, those heavy elements would fluoresce under an X-ray.

That's a big "if." Ancient Egyptians used a variety of writing utensils. To an X-ray, a carbon-based ink would blend in with the carbon-based surroundings of the papyrus—the whole thing would light up.

At the Lawrence Berkeley Lab is the Advanced Light Source, a facility funded by the US Department of Energy. It provides researchers in a variety of disciplines the use of a blend of soft X-rays, hard X-rays, and infrared light. And the images it produces have 3-micron resolution—that's tight and clear enough to show not the ink, which may blend in with the paper, but the impressions a writing instrument made while being pressed against the papyrus. It's like reading a grocery list by reading the impressions left on the sheet of paper under the one on which the list was written.

The researchers who met at the Lawrence Berkeley Lab are starting their investigation slowly (the data from which they will make available for free online). They draw on blank pieces of papyrus with a variety of paints, pens, and pencils and then roll them, crumple them, and layer them, image side in. We have so many technologies to try but precious few ancient pieces of cartonnage.

REVEALING ANCIENT WORDS

Salvaging Burnt Paper

X-rays can even make burnt paper readable again! During the devastating eruption of Mount Vesuvius in 79 CE, nearly everything in the path of the floods of lava and the bursts of scorching-hot gas got burned. Hundreds of Herculaneum scrolls, named later for the city in which they were found, were carbonized in the 608 degree Fahrenheit (320 degree Celsius) heat of the volcano's blast. These tightly rolled scrolls now look like tree stumps of coal, but each may be lengths of paper 49 feet (15 m) long—that's a lot of precious insight from one of humanity's ancient societies.

Scientists at the National Research Council in Naples, Italy, developed X-ray phase contrast tomography (XPCT). The technology can, astoundingly, pick out black ink against the black char. This is not to say the words are easy to read. Researchers have mostly been able to read a letter here and there; they have been able to decipher only two whole words, one meaning "would fall" and one meaning "would say." But they do have enough evidence to think the handwriting matches that of first-century BCE philosopher Philodemus. A machine called a **synchrotron** may be able to produce clearer images.

Finding Hidden Manuscripts

Bookbinders from the fifteenth through eighteenth centuries recycled, using pieces of handwritten manuscripts, which were discarded as old-fashioned after the widespread use of the printing press. They used these pieces to strengthen the bindings of new volumes.

Thousands of those books exist in every great library in the world, so there are lots of places for scraps of older manuscripts to be hiding. Because researchers don't want to destroy those bindings, and don't know which part of which

Abigail Quandt, museum conservator, uses X-rays at Stanford University's Linear Accelerator Center to decipher hidden writings of Archimedes.

bindings might contain such reinforcement anyway, they needed a technology a little more advanced than a knife. Enter the X-ray: a thin beam scans the object (slowly—sometimes taking more than twenty-four hours per scan), reacting to the different elements it meets. In this way, the iron, copper, and zinc of medieval ink stand out even when hidden from normal view by heavy parchment.

REVEALING DISCARDED PERSPECTIVES

X-ray fluorescence spectrometry was originally developed for use on paintings. X-rays can reveal paintings underneath paintings. Under the final product that hangs in a museum may be the artist's messy first draft that was covered over to save the cost of a new canvas or the controversial view the artist wished to express but changed before the final showing. X-rays have revealed that Picasso's *The Old Guitarist*, painted during his Blue Period of 1901–1904, covers the image of a woman nursing a child, a bull and sheep grazing nearby. John Singer Sargent painted *Madame X* (1883–1884), a society woman who had moved from New Orleans to Europe to make a name for herself, with one jeweled strap of her gown slipped from her shoulder. Upon its first public showing, people were angry about how scandalous that made her look, and her embarrassed family asked for it to be removed. Sargent repainted the strap tightly set over her shoulder. That this happened was no secret, but with X-ray technology, we can now see what was.

X-rays of paintings help with restorations and can also prove forgeries—a forger may accurately copy the final work, but did he or she remember that that artist usually produced what would become an underpainting? Did the forger match the quality of his or her underpainting with that of the real artist?

REVEALING MORE ANCIENT LIVES

Trapped in Amber

Just as X-rays help Egyptologists look at and through cartonnage without destroying it, they also help paleontologists see the bodies of fossilized creatures without breaking open the stones that preserved them. In 2009, paleontologists from the European Synchrotron Research Facility in France used brand-new high-energy X-rays, accelerator technology, to examine what lay inside 640 pieces of opaque amber from 145 to 65 million years ago. Most of the amber from that period in that area of the world is opaque. While fossilized tree sap is usually transparent, if dirt or other debris settles in it, it becomes as opaque as rock; no naked human eye can see through that. Past viewing of archaeological finds used CT scanners like those used in hospitals—not strong enough to visually bust through opaque amber. With this new form of X-ray, we have met 356 new mites, wasps, flies, spiders, and plants.

First, the team aimed a synchrotron X-ray beam at the amber. By noting what spots modified the wavelength shapes, researchers could see where the fossils of plants or animals were in the amber. Second, they then aimed the beam at amber as the piece rotated, in order to create a 3D rendering of the creature within.

These synchrotron-based X-rays are helping scientists to focus on each individual chemical component of a fossil. For example, researchers could focus on calcium alone. This will allow scientists to reconstruct the soft tissues of ancient animals. Remnants of the chemical elements of some of those remain in the fossils. By "seeing" them with X-rays, we can re-create them.

Archaeopteryx

Synchrotrons have helped scientists to make a lot of discoveries about our world's distant past. In 2010, researchers using the Stanford Synchotron Radiation Lightsource (SSRL) made an amazing find: the chemical makeup of the "dinobird."

The *Archaeopteryx* fossil proves a chemical link between dinosaurs and present-day birds.

Archaeopteryx was discovered at a key point in history: only one year after the publication of Charles Darwin's *On the Origin of Species*, this paleontological find of a half-dinosaur/half-bird offered strong proof of Darwin's theory of evolution. One hundred fifty years and ten *Archaeopteryx* specimens later, scientists have found something new in this fossil.

The scientists passed the synchrotron's hair-thin X-ray beam across the 150-million-year-old Thermopolis *Archaeopteryx* specimen. The X-ray reflected differently over each component it struck, revealing not only that chemical residue remained but which chemicals and where they were in the bones. Though we knew before that dinosaurs and birds are physically linked, we now know they are chemically linked as well—the same chemicals found in *Archaeopteryx* are also found in today's birds.

This will affect all of paleontology because scientists will now look at bones as doing more than "just" showing an ancient creature's skeleton; they may contain chemical clues. This could lead to advancements in technology. X-raying fossils for chemicals is best done with the fossil still embedded in stone—that helps paleontologists confirm that a trace chemical belongs to the animal and not to the rock. Perhaps smaller machines will be invented so that they can be carried right to the excavation site.

NOVELTY TURNING INTO NECESSITY

X-ray glasses of popular culture—those ones with squiggly lines over the lenses that are sometimes given away as toys in cereal boxes—don't really allow the wearer to see through things. They offer an optical illusion only. But in 2014, a company created real X-ray glasses! Eyes-On glasses could be of great use to **phlebotomists**—and the patients their needles poke. Wearing them, phlebotomists could see through skin and find by sight the strongest vein from which to draw blood.

TECHNOLOGY IS USELESS IF IT CAN'T BE USED

Ultimately, however, if people don't have access to X-rays, for financial or geographic reasons, the technology does them no good. And that's a serious problem in our modern world. Even minor dental problems can lead to major health issues, including death. An untreated cavity can lead to infection, which can move into bone or blood, causing sepsis that affects major organs.

Financial Barriers to Dental X-rays

Though geography can pose a challenge for Americans to see a dentist (consider the people living on remote islands of Alaska, the communities deep within the Appalachian Mountains, or even those who live in the Midwest farmland an hour's drive from a town of any size), much of why people in the United States don't see a dentist is related to cost. *A Costly Dental Destination* (2012), a Pew Charitable Trust study, estimates "that preventable dental conditions were the primary reason for 830,590 ER visits by Americans in 2009—a 16 percent increase from 2006." Researchers of a study published in *Journal of Endodontics* (2013) found that between 2000 and 2008, there were more than sixty-one thousand hospitalizations across the United States for periapical abscesses, an infection that is both a common symptom of untreated tooth decay and commonly spotted using images from periapical X-rays. Sixty-six of the people hospitalized with that infection died.

Geographic Barriers to Dental X-rays

In 2013, it was reported that there was one dentist for every one hundred thousand people in Tanzania. In comparison, there was one dentist for every twenty-five hundred people in the United Kingdom. In Rwanda, there were eleven dentists total. Most of these professionals are clustered in urban centers, whereas most people in those African countries live in rural areas that are both far from the cities and without access to good roads, even if people had transportation.

Financial and Geographic Barriers to Medical X-rays

Accessibility remains a challenge to general medical care in many parts of the world, too. X-rays have been known of and used for more than one hundred years, but that doesn't mean that every person who needs them has the option of receiving that help.

Some aren't x-rayed due to financial or geographic obstacles. In 2011, bonesetters still provided the majority (70–90 percent) of bone fracture care in certain areas of Nigeria. The country had only three major centers devoted to orthopedics and fewer than two hundred orthopedic surgeons. The National Orthopaedic Hospital Enugu, researched for a study published in the *Open Orthopaedics Journal*, is one of those three, and even it did not have diagnostic resources such as MRI or CT.

Prejudicial Barriers to X-rays

Even with access to X-rays, not everyone will choose them. Some of the reason people don't get X-rays is because they fear the technology, don't feel familiar with medical centers, or believe strongly in the powers and abilities of alternative health care providers. The world's history of racial and economic inequality also plays a role in why someone might not choose a technology like X-rays. In Nigeria, for example, the first hospital opened in 1873, but it was founded by and for

European colonialists. Nigerians weren't allowed to receive its services for another forty years, and then only the most elite did. Nigerian physicians first received training in Western orthopedics once the country declared its independence from Great Britain—in 1960, almost one hundred years after the first hospital opened there. After such a segregated past, it stands to reason that people, patients and medical professionals alike, would be slow to integrate Western and traditional techniques.

X-RAYS and TERRORISM

If you were playing a game in which you had to name the most common places X-rays are used, the medical world would surely be one of your immediate answers; airport travel would likely be the other.

Attempts to Thwart Terrorism

Before 9/11, air travel was very different from how it is now, including what you could pack in your carry-on luggage or even on your body. Before 2001, you could carry on to a plane baseball bats, box cutters, even knives up to 4 inches (10.2 cm) long. Two months after the 9/11 attacks, under the Aviation and Transportation Security Act, airport safety transitioned from a job for private security companies to a job of the federal government, specifically, the new Transportation Security Administration.

The TSA slowly rolled out full-body X-ray scanners to airports around the United States, with the technology becoming commonplace by 2010. All airline passengers must either go through the scanner or get a full-body pat down. Before the heightened security at airports, passengers walked through metal detectors meant to detect big weapons, like guns, not small but still potentially harmful ones, like those

X-rays and Airports: Privacy, Safety, and Health

Changes had to be made in airport travel after 9/11. It would have been unacceptable for the government to continue allowing passengers to stroll relatively unchecked from the outside world onto planes. The nation needed to be assured that terrorists masquerading as passengers could never again take down a plane from within the plane. But from the start, many have doubted the efficacy of the Transportation Security Administration and its full-body scanners, the first of which were X-ray machines.

A leaked report from May 2015 showed that fake explosives and guns, planted to test TSA employees, were undetected by screeners 95 percent of the time. On August 10, 2015, a man who'd flown from Detroit, Michigan, to Chicago, Illinois, turned himself in to TSA because while on the flight he realized he'd forgotten to remove a prohibited item from his bag, a revolver and bullets, which X-ray scanners and the screeners monitoring them in Detroit had completely missed.

Scientists urged for more research to be done on the X-ray machines to confirm they were not detrimental to passengers' health. Both when X-ray-emitting machines were used and after they were switched to those using radio waves, the public worried about being exposed to cancer-causing radiation.

People have pushed back against the full-body photo-imaging scanners since it became apparent

Travelers at Moscow's Domodedovo Airport step through X-ray machines.

they weren't a passing fad and were, in fact, going to become more and more prevalent. Activists named November 24, 2010, National Opt-Out Day, claiming scanners made air travelers guilty until proved innocent, and privacy groups started filing court orders calling the scanners illegally invasive. Though most people have always been allowed to opt-out of the scanners, the alternative has been extensive pat downs by TSA employees.

small knives mentioned before. The metal detectors weren't strong enough to sense those. X-ray scanners allow TSA officials to spot metallic and nonmetallic potential weapons, such as plastic explosives.

The first airport scanners were backscatter X-rays. Steven W. Smith patented the airport prototype and then sold it to what would become one of the three companies that manufactured them. Smith explained that a "pencil beam" scans the body; the rays scattered from the body create an image of the person—and anything the person is carrying. Heavy objects, like metal, backscatter electrons more strongly than light objects, like organic materials and plastics. Therefore, heavy objects show up darker in the produced image than the light objects. The staff person monitoring the X-ray images can see the dark shape of a knife standing out against a light hipbone, for example.

Airplane passengers had major issues with these first-generation X-ray machines for two big reasons. Backscatter also showed rather graphic images of passengers, detailing not only everything they carried but their body parts themselves. As an X-ray machine, it emitted ionizing radiation, albeit at small levels, at each person who stood in the machine. Both of these things concerned people. By May 2013, all of these machines had been replaced with Advanced Imaging Technology, which uses millimeter waves, the kind of radio waves that cell phones give off. And they show each passenger as the same gingerbread-cookie shape, with suspicious items determined not by staring TSA staff but by a computer program that highlights the objects with yellow boxes.

Attempts to Terrorize

As the TSA is attempting to use X-rays to prevent acts of terrorism, there have been reports of people trying to use X-rays to incite terror. In 2013, two New Yorkers developed a remote-controlled device that would send out dangerous levels of X-ray radiation. They were planning to aim the beams at members of religious groups the attackers found "undesirable." Eight to ten grays (a unit of radiation dose) from the device, which would have fit in a truck and been powered by the truck's built-in cigarette lighter, wouldn't kill a target immediately, but it would soon. The men were arrested and charged with conspiracy.

FOOD SAFETY

Looking for Unintended Ingredients

The food-processing industry has long put food through metal detectors to check for any bits of assembly-line machinery that may have loosened and fallen into products. Federal rules state that food has to be reviewed for quality at "hazard analysis critical control points," those moments that are ripe for error. For example, a hazard analysis questionnaire asks if there are any sensitive ingredients in the food being processed that may present issues; if water is being used in processing the food, the quality of which may affect the food; and if there's the possibility of contamination between processing and packaging. In the 2000s, the industry made a leap forward by adding X-ray scanners to the protocol in order to see a greater variety of foreign objects in a greater variety of products.

These machines are more high-tech versions of the ones that scan luggage at the airport. They scan multiple items each second, with software, not human eyes, monitoring the results. A computer learns what's considered good and what's considered unacceptable by trainers showing it clean food and then contaminated food and allowing it to record the differences for comparison against future products.

Metal detectors cannot sense plastic, bone, and rock, but that kind of debris is possible in food. X-rays can detect these things, though. Fruit pies might contain bits of rock, inadvertently brought along from the harvest. Or consider hot dogs—bone and hard cartilage fragments easily sneak through processing into the final product. Costco was one of the big retailers being told by its customers that there were problems. The company's vice president of quality assurance and food safety told National Public Radio that under the system of metal detection only, he heard about fifteen to twenty broken teeth a month after people chomped into unexpected bone in hot dogs. Since the company used its own purchasing power in 2012 to persuade the hot dog manufacturer to add X-ray machines, it's received no reports of broken teeth.

Metal detectors can't read metal particulates in certain products. Wet and salty foods, from blocks of cheese to frozen pizza, are tricky. And, of course, the metal containers of some foods, such as pies and soft drinks, mask any metal tagging along inside the food. Since 2005, the manufacturer of pies for the Bakers Square restaurant chain has used X-ray machines.

Manufacturers, who are considered 100 percent liable under most states' laws if their products cause issues for consumers, see the huge price tag for the machines—up to $70,000 in 2012—as worth it. And research says the X-rays don't create radioactive food; they operate at the level of power a 100-watt lightbulb does, much lower than medical X-rays and much lower than those that are used to kill microbes in food.

Gamma radiation killed microorganisms on and in the strawberries on the left, prolonging their shelf life.

Killing Food-Borne Illnesses

Not even food irradiation (treating food with a particular kind of radiation) makes food radioactive, says the Food and Drug Administration (FDA). Irradiation disappears when its energy source, such as X-rays, is removed. The FDA has evaluated the safety of irradiated food for thirty years and now regulates the zapping of meat, seafood, and fresh fruits and vegetables with gamma rays, electron beams, or, since 2009, X-rays. These methods prevent food-borne illness, preserve food, control insects, delay food aging, and sterilize.

Roughly seventy-six million Americans get sick each year from food-borne illnesses. Some even die from products tainted with salmonella and *E. coli*. Irradiation eliminates these organisms. It also halts the microorganisms that hasten decomposition and slows processes like potatoes sprouting and fruit ripening, extending the shelf life of food. Irradiation destroys insects and can be used to sterilize food. Sterilization requires a higher dosage of treatment than most food is cleared to receive, so it's used only on products that will go to places such as hospitals, for patients with severely impaired immune systems, and space, for astronauts who can't afford to get sick so far from home.

Also known as cold pasteurization, irradiation advanced in 2009 with the study of X-rays' efficacy on food. People had been considering X-rays for this purpose for a while, but only four commercial X-ray irradiation units were built between 1996 and 2009. Then Michigan State University investigators determined that a dose of X-rays lower than the dose needed by gamma rays killed 99.999 percent of pathogens.

To determine what kind of X-ray dose each unique food needed, investigators injected the food with bacteria, put it in the X-ray machine (the prototype was the size of three refrigerators put together), and then counted the number of bacteria that survived the irradiation and checked if the food's physical state had changed.

The World Health Organization (WHO), the Centers for Disease Control and Prevention (CDC), and the US Department of Agriculture (USDA) have joined the FDA in approving irradiated food. Consumers can tell that the food they're about to buy has been irradiated if the FDA-required symbol and words "Treated with radiation" or "Treated by irradiation" are present on the nutrition label. Regulators of food sold as organic do not allow irradiation. The Organic Consumers Association says irradiated food has depleted vitamins and enzymes and that the labeling is not as accurate as it needs to be. After organic California spinach tainted with *E. coli* made at least 205 people sick and killed three, the packager instituted a process of triple-washing food with chlorine and testing for pathogens throughout production. By 2008, 0.14 percent of the greens that arrived at the plant tested positive for salmonella or *E. coli*.

RAYS in SPACE

X-ray Astronomy

The space race has long been tied to the war industry, and it was through use of a captured German missile in 1949 that scientists created X-ray astronomy. They attached small **Geiger counters** to the missile and sent it 50 miles (80.4 km) up, from where it reported back that the sun (and therefore all stars) emits X-rays. We couldn't have known that before—not only because it hadn't been long since the reveal of X-rays but also because Earth's atmosphere absorbs X-rays. We needed to be able to reach space ourselves in order to understand what other celestial objects were outputting. As the experiments grew in number, tracing the paths of X-rays through outer space has revealed the existence, locations, and star populations of distant galaxies. The presence of X-rays also points to the location of black holes. As matter swirls around a black hole, being sucked in a fast-moving spiral toward its center, heat is created; the area gets so hot, it mimics a star's heat and produces X-rays.

Chandra

On July 23, 1999, NASA launched the Chandra X-ray Observatory into space aboard the *Columbia*; Chandra has been orbiting 86,500 miles (139,208.3 km) above Earth ever since. As of 2014, it was considered one of the most successful NASA missions ever, outperforming its schedule and technical expectations and delivering so much information to scientists back on Earth that its cost has been a bargain. One of its big accomplishments has been to send us information about X-rays perhaps coming from dark matter. Astronomers think dark matter makes up the vast majority of the universe, but it's very difficult to study.

The UNKNOWN THAT HAS DONE SO MUCH

The "X" in "X-ray" stands for "an unknown quantity." X-rays were the mysterious rays that no one was looking for but that we have used to do so much good. They're time-traveling devices of a sort, showing us glimpses of what happened in the past, informing us in the present, and helping us to imagine our future. Thanks to X-rays, paleopathologists can see disease in the bones of ancient humans and archaeologists can learn about ancient civilizations by seeing through mummy wrappings and reading burnt documents. Today's doctors and dentists are able to diagnose illnesses earlier than ever before and pinpoint the locations of injuries, and food processors are able to check inside food for unwanted debris. NASA engineers receive X-rays from galaxies far away, helping us understand where our exploration of the universe may take us in the future.

X-rays also produce incredibly dangerous levels of radiation, which can harm or kill healthy human cells as well as they can unhealthy ones. And X-rays are not the miracle answer to our problems. Even though we sometimes treat them as such, they are a tool. Consider the full-body X-ray scanners in airport security. Alone, X-rays don't catch illegal items in luggage; we rely on human interpretation of X-ray images to understand what's there.

X-rays are one of the most incredible accidental revelations. Scientists will continue to devise new ways to use them—and if they take their cue from Röntgen, they'll keep their minds open to the possibilities of other fantastic happenings in the lab.

Chronology

5000 BCE First recorded mention of dental decay

3000 BCE The Edwin Smith Papyrus, the first known medical textbook, is written

700 BCE Bones from this era show signs of prostate cancer

400s–50 BCE First use of the word "cancer"

255–206 BCE First cleft lip repair

1210 Guild of Barbers is established in France

Early 1600s Chinese dental surgeons treat tonsillar abscesses to eptheliomas of the lip

Early 1600s The London Bills of Mortality lists "Teeth" as the fifth or sixth most common cause of death

1657 The Magdeburg hemispheres experiment takes place

1700s Most people see bonesetters, not doctors, for broken bones

1705 Electrical discharges are found to travel a longer distance in vacuums than outside them

1800 The battery is invented

1820 The connection between electricity and magnetism is demonstrated

1832 William Crookes is born

1844 Samuel Morse sends the first telegraph message

1846 First surgery with anesthetic takes place

1866 First permanent telegraph cable is laid across the Atlantic Ocean, connecting the United States and Europe; there are forty transatlantic lines by 1940

1881 James A. Garfield is shot, and he dies eighty days later

1882 First radical mastectomy

1895 Wilhelm Röntgen discovers X-rays

1904 American Lung Foundation forms to combat tuberculosis

2001 September 11 terrorist attacks; subsequent changes in airport security lead to the installation of full-body X-ray scanners for passengers

2008 The novel the *Bonesetter's Daughter* becomes an opera

Glossary

anesthetic A drug that numbs the body and keeps a person from feeling pain; used during surgery.

anode The positively charged electrode; electrons leave a device through this.

bacterium A unicellular microorganism that can cause disease.

carcinogen Something that can cause cancer.

cathode The negatively charged electrode; electrons enter a device through this.

Crookes tube Named for English physicist William Crookes, one of many who worked on creating such a device around 1869 to 1875; a discharge tube with a vacuum in which cathode rays, and then X-rays, were discovered.

crystallography A science that studies the composition of substances by looking at their crystallized forms.

CT scan An X-ray procedure that produces detailed, cross-sectional images of the body.

disease A condition that prevents normal function of a part of or whole living creature or plant; often includes distinguishing symptoms.

electromagnetism A branch of physics that studies the interaction of electrically charged particles.

evacuated Describes a container devoid of air.

fluoroscope A device with a fluorescent screen for viewing X-rays in real time, not via X-ray photograph.

Geiger counter A radiation detector that detects the emissions of a wide range of particles and rays, including X-rays.

germ A microorganism that causes disease.

miasma theory An early theory that vapor or air caused disease.

micro-CT Microtomography; X-ray imaging in 3D.

orthopedics The branch of medicine that deals with bones.

paleopathologist Someone who studies ancient diseases or diseases found in remains of ancient humans and animals.

phlebotomist Someone who draws blood from patients.

radiograph An image produced by X-rays.

radiologist Medical doctor who diagnoses and treats diseases and injuries using imaging techniques like X-rays.

shadowgraph An image produced by X-rays.

spiritualism Belief based on communication with the dead.

synchrotron An extremely powerful source of X-rays.

vacuum pump A device that removes gas or air from a container, leaving behind a partial vacuum.

Further Information

BOOKS

Fadiman, Anne. *The Spirit Catches You and You Fall Down: A Hmong Child, Her American Doctors, and the Collision of Two Cultures.* New York, NY: Farrar, Straus and Giroux, 2012.

Veasey, Nick. *X-treme X-ray: See the World Inside Out!* New York, NY: Scholastic Paperbacks, 2010.

WEBSITES

Science: X-rays
http://www.bbc.co.uk/schools/gcsebitesize/science/triple_edexcel/xrays_ecgs/xrays/revision/1/

Explore pictures and a roundup of facts about the medical uses of X-rays.

Who Invented the X-ray?
http://science.howstuffworks.com/innovation/inventions/who-invented-the-x-ray.htm

Learn more about the discovery of X-rays on this website from How Stuff Works.

X-rays
https://www.nobelprize.org/educational/physics/x-rays/index.html

The Nobel Foundation provides a comprehensive look at the discovery of X-rays and how they work.

VIDEOS

"115-Year-Old Medical X-ray Machine Comes Back to Life"
https://www.wired.com/2016/10/x-rays-revealing-mysterious-writings-mummy-coffins/

Watch one of the first X-ray machines at work. The look and sound of it as well as how incredibly long the woman must hold her hand in it are fascinating.

"Revealing Letters in Rolled Herculaneum Papyri by X-ray Phase-Contrast Imaging"
https://embed.theguardian.com/embed/video/science/video/2015/jan/20/herculaneum-texts-vesuvius-x-ray-imaging-video

As this video shows, X-rays don't eliminate the need for the trained eyes and brains of expert researchers. It takes a lot of skill to decipher what the X-ray reveals.

"X-rays Are Revealing the Mysterious Writings in Mummy Coffins"
https://www.wired.com/2016/10/x-rays-revealing-mysterious-writings-mummy-coffins/

Watch how difficult it can be to examine archaeological finds without X-rays.

Bibliography

Alberge, Dalya. "X-rays Reveal 1,300-Year-Old Writings Inside Later Bookbindings." *Guardian*, June 4, 2016. https://www.theguardian.com/books/2016/jun/04/x-rays-reveal-medieval-manuscripts.

Alexander, Steve. "Seeing Through the Pie Crust." *Star Tribune*, January 7, 2012. http://www.startribune.com/seeing-through-the-pie-crust/136812378/.

Assmus, Alexi. "Early History of X Rays." *Beam Line*, Summer 1995. http://www.slac.stanford.edu/pubs/beamline/25/2/25-2-assmus.pdf.

Associated Press. "Just What Can They See?! Your Full Body Scanner Questions Answered." *New York Daily News*, December 31, 2009. http://www.nydailynews.com/news/money/full-body-scanner-questions-answered-article-1.434194.

Barber, Gregory. "X-rays Are Revealing the Mysterious Writings in Mummy Coffins." *Wired*, October 26, 2016. https://www.wired.com/2016/10/x-rays-revealing-mysterious-writings-mummy-coffins/.

Bloxham, Andy. "X-ray Voted Most Important Modern Discovery." *Guardian*, November 4, 2009. http://www.telegraph.co.uk/news/science/science-news/6498941/X-ray-voted-most-important-modern-discovery.html.

Bradford, Isabella. "Mending Broken Bones, c. 1769." *Two Nerdy History Girls* (blog), February 26, 2014. http://twonerdyhistorygirls.blogspot.com/2014/02/mending-broken-bones-c-1769.html.

Brorson, Stig. "Management of Fractures of the Humerus in Ancient Egypt, Greece, and Rome: An Historical Review." *Clinical Orthopaedics and Related Research* 467 (2009), 1907–1914. https://www.ncbi.nlm.nih.gov/pmc/articles/PMC2690737/.

Brown, Percy. "Clarence Madison Dally (1865–1904)." *AJR* 164 (1995): 237–239.

Brundtland, Terje. "Francis Hauksbee and His Air Pump." The Royal Society Publishing, July 11, 2012. http://rsnr.royalsocietypublishing.org/content/66/3/253.

Bynum, Helen. *Spitting Blood: The History of Tuberculosis.* Oxford, UK: Oxford University Press, 2012.

Chicago Radiological Society. "Brief Biography of Emil Grubbe." Retrieved December 26, 2016. http://www.chi-rad-soc.org/grubbe.htm.

Cyran, Pamela, and Chris Gaylord. "The 20 Most Fascinating Accidental Inventions." *Christian Science Monitor*, October 5, 2012. http://www.csmonitor.com/Technology/2012/1005/The-20-most-fascinating-accidental-inventions/X-ray-images.

Daily Mail. "Google 'Doodle' Celebrates 115 Years Since Discovery of X-rays Plus Nod to 'PigeonRank' April Fool's Joke." November 10, 2010. http://www.dailymail.co.uk/sciencetech/article-1327641/X-Rays-Google-doodle-celebrates-115-years.html.

Diamond, Giselle. "How Do Vacuum Pumps Work?" eHow.com. Retrieved December 14, 2016. http://www.ehow.com/how-does_5016448_vacuum-pumps-work.html.

Encyclopaedia Britannica. "Sir William Crookes." October 24, 2003. https://www.britannica.com/biography/William-Crookes.

———. "Wilhelm Conrad Röntgen." December 29, 2011. https://www.britannica.com/biography/Wilhelm-Rontgen.

Ferrigno, Lorenzo. "Feds Nab KKK Member, Accomplice for Lethal X-ray Plot." CNN, June 21, 2013. http://www.cnn.com/2013/06/20/justice/new-york-x-ray-plot/.

Frame, Paul W. "Wilhelm Röntgen and the Invisible Light." ORAU. Retrieved December 26, 2016. https://www.orau.org/ptp/articlesstories/invisiblelight.htm.

Gerson, Edwin S. "X-ray Mania: The X Ray in Advertising, Circa 1895." *RadioGraphics* 24 (March 2004). http://pubs.rsna.org/doi/full/10.1148/rg.242035157.

Gillin, Joshua. "Dentists and the ADA Agree: Ignoring a Toothache Could Potentially Kill You." *Politifact*, January 26, 2016. http://www.politifact.com/florida/statements/2016/jan/26/john-cortes/ignoring-toothache-really-could-potentially-kill-y/.

Grossman, Lisa. "115-Year-Old Medical X-ray Machine Comes Back to Life." *Wired*, March 16, 2011. https://www.wired.com/2011/03/old-x-rays/.

Gunderman, Richard. "On the 120th Anniversary of the X-ray, a Look at How It Changed Our View of the World." The Conversation, November 6, 2015. http://theconversation.com/on-the-120th-anniversary-of-the-x-ray-a-look-at-how-it-changed-our-view-of-the-world-50154.

Hallyn, Fernand, ed. *Metaphor and Analogy in the Sciences.* Dordrecht, Netherlands: Springer Science + Business Media. https://books.google.com/books?id=iKleTxffj5IC.

History.com. "Morse Code & the Telegraph." 2009. http://www.history.com/topics/inventions/telegraph.

Johnson, Leland R. "Ernest Omar Wollan." The Tennessee Encyclopedia of History and Culture. Last updated January 1, 2010. http://tennesseeencyclopedia.net/entry.php?rec=1527.

Jones, Andrew Zimmerman. "Cathode Ray." About.com, October 3, 2016. http://physics.about.com/od/glossary/g/cathoderay.htm.

Kelley. "A Short Painful History of Dentistry." HubPages. Last updated November 10, 2016. http://hubpages.com/health/A-Short-Painful-History-of-Dentistry.

Kirsh, Andrea, and Rustin S. Levenson. *Seeing Through Paintings: Physical Examination in Art Historical Studies.* New Haven, CT: Yale University Press, 2002. https://books.google.com/books/about/Seeing_Through_Paintings.html?id=fB_LlpuB4gsC.

Kopp, H. "Radiation Damage Caused by Shoe-Fitting Fluoroscope." *British Medical Journal* (December 7, 1957): 1344–1345.

Kuruvilla, Carol. "TSA Has Completely Removed Revealing X-ray Scanners: Rep." *New York Daily News*, May 31, 2013. http://www.nydailynews.com/news/national/tsa-completely-removed-full-body-scanners-rep-article-1.1360143.

Limer, Eric. "The Insane Cancer Machines that Used to Live in Shoe Stores Everywhere." *Gizmodo*, July 15, 2013. http://gizmodo.com/the-insane-cancer-machines-that-used-to-live-in-shoe-st-789073694.

Lovejoy, Bess. "9 Transparently Amazing Facts About X-rays." *Mental Floss*, November 7, 2015. http://mentalfloss.com/article/70900/9-transparently-amazing-facts-about-x-rays.

Mandal, Ananya. "History of Breast Cancer." News-Medical.net. Last updated September 22, 2013. http://www.news-medical.net/health/History-of-Breast-Cancer.aspx.

———. "Cancer History." News-Medical.net. Last updated February 21, 2014. http://www.news-medical.net/health/Cancer-History.aspx.

Markel, Howard. "How Playing with Dangerous X-rays Led to the Discovery of Radiation Treatment for Cancer." *Newshour*, January 28, 2015. http://www.pbs.org/newshour/updates/emil-grubbe-first-use-radiation-treat-breast-cancer/.

Neuhauser, Alan. "15 Years After 9/11, TSA Is Still Falling Short." *U.S. News & World Report*, September 9, 2016. http://www.usnews.com/news/articles/2016-09-09/15-years-after-9-11-the-tsa-is-still-falling-short.

Nwachukwu, Benedict U., Ikechukwu C. Okwesili, Mitchell B. Harris, and Jeffrey N. Katz. "Traditional Bonesetters and Contemporary Orthopaedic Fracture Care in a Developing Nation: Historical Aspects, Contemporary Status and Future Directions." *Open Orthopaedics Journal* 5 (2011): 20–26. https://www.ncbi.nlm.nih.gov/pmc/articles/PMC3027080/.

O'Connor, Lydia. "This Is What It Was Like to Go to the Airport Before 9/11." *Huffington Post*, September 11, 2016. http://www.huffingtonpost.com/entry/airports-before-911_us_57c85e17e4b078581f11a133.

Pendergrass, Eugene P. "Book Review: The Life and Times of Emil H. Grubbe by Paul C. Hodges." *Isis: A Journal of the History of Science Society* 56 (Autumn 1965). http://www.journals.uchicago.edu/doi/10.1086/350038.

Pew Charitable Trusts. "A Costly Dental Destination." February 28, 2012. http://www.pewtrusts.org/en/research-and-analysis/reports/2012/02/28/a-costly-dental-destination.

Rocca, Mo. "How Doctors Killed President Garfield." CBS News, July 5, 2012. http://www.cbsnews.com/news/how-doctors-killed-president-garfield/.

Sakorafas, G. H., and M. Safioleas. "Breast Cancer Surgery: An Historical Narrative. Part I. From Prehistoric Times to Renaissance." *European Journal of Cancer Care* 18 (6) (November 2009): 530–544. https://www.ncbi.nlm.nih.gov/pubmed/19674074.

Sample, Ian. "Words Emerge from Ancient Scrolls Charred During Eruption of Vesuvius." *Guardian*, January 20, 2015. https://www.theguardian.com/science/2015/jan/20/words-ancient-scrolls-eruption-vesuvius-x-ray-herculaneum.

Sansare, K., V. Khanna, and F. Karjodkar. "Early Victims of X-rays: A Tribute and Current Perception." *DentoMaxilloFacial Radiology: A Journal of Head & Neck Imaging* 40 (February 2011): 123–125.

Schaffer, Amanda. "A President Felled by an Assassin and 1880's Medical Care." *New York Times*, July 25, 2006. http://www.nytimes.com/2006/07/25/health/25garf.html.

Schmidt, Fabian. "X-ray Vision: An Accidental Discovery That Revolutionized Medicine." DW, November 6, 2015. http://www.dw.com/en/x-ray-vision-an-accidental-discovery-that-revolutionized-medicine/a-18833060.

Seliger, Howard H. "Wilhelm Conrad Röntgen and the Glimmer of Light." *Physics Today* (November 1995): 25–31. https://www.physics.rutgers.edu/grad/612/seliger_nov95.pdf.

Shute, Nancy. "X-rays Scan Foods for the Secret Ingredient That Could Break a Tooth." NPR, January 10, 2012. http://www.npr.org/sections/thesalt/2012/01/10/144957614/x-rays-scan-foods-for-the-secret-ingredient-that-could-break-a-tooth.

Skolnick, Andrew A. "Natasha Demkina: The Girl with Very Normal Eyes." Live Science, January 28, 2005. http://www.livescience.com/109-natasha-demkina-girl-normal-eyes.html.

Smullen, Michael J., and David E. Bertler. "Basal Cell Carcinoma of the Sole: Possible Association with the Shoe-Fitting Fluoroscope." *Wisconsin Medical Journal* 106 (2007): 275–278.

Spiegel, Peter K. "The First Clinical X-ray Made in America—100 Years." *AJR*, 164 (1995): 241–243. http://www.ajronline.org/doi/pdf/10.2214/ajr.164.1.7998549.

Starr, Michelle. "CT Scan Finds Mummified Monk Inside 1,000-Year-Old Buddha." *CNet*, February 22, 2015. https://www.cnet.com/news/ct-scan-finds-mummified-monk-inside-1000-year-old-buddha/.

Strauss, Mark. "Earliest Human Cancer Found in 1.7-Million-Year-Old Bone." *National Geographic*, July 28, 2016. http://news.nationalgeographic.com/2016/07/oldest-human-cancer-disease-origins-tumor-fossil-science/.

Swallow, Erica. "The Science Behind Airport Body Scanners." *Mashable*, November 17, 2011. http://mashable.com/2011/11/17/tsa-body-scanner/.

Than, Ker. "Hundreds of Dino-Era Animals in Amber Revealed by X-ray." *National Geographic News*, April 4, 2008. http://news.nationalgeographic.com/news/2008/04/080404-amber-animals.html.

———. "X-rays on Google: Surprising Ways the Rays Are Used Today." *National Geographic News*, November 10, 2010. http://news.nationalgeographic.com/news/2010/101108-x-rays-google-doodle-115th-anniversary-years-science-logo/.

Topley, Mark. "Surely People Don't Die from a Toothache." Oral Health Foundation, March 20, 2013. https://www.dentalhealth.org/blog/blogdetails/70.

Voakes, Greg. "The First X-ray Technician: 4 Weird Facts About X-ray Inventor Wilhelm Conrad Roentgen." *Visual News*, August 19, 2011. https://www.visualnews.com/2011/08/19/first-x-ray-technician-4-weird-facts-about-x-ray-inventor-wilhelm-conrad-roentgen/.

Waters, Hannah. "The First X-ray, 1895." *The Scientist*, July 1, 2011. http://www.the-scientist.com/?articles.view/articleNo/30693/title/The-First-X-ray--1895/.

Wayland, Greg. "Inventor of X-ray Technology Behind Airport Scanners Speaks Out." NECN, March 25, 2014. http://www.necn.com/news/new-england/_NECN__Inventor_of_X-ray_Technology_Behind_Airport_Scanners_Speaks_Out_NECN-252291351.html.

Zetter, Kim. "National Opt-Out Day Called Against Invasive Body Scanners." *Wired*, November 12, 2010. https://www.wired.com/2010/11/national-opt-out/.

Zorich, Zach. "Fixing Ancient Toothaches." *Archaeology*, December 19, 2012. http://www.archaeology.org/issues/60-1301/trenches/321-lonche-tooth-beeswax-dentistry.

Index

Page numbers in **boldface** are illustrations. Entries in **boldface** are glossary terms.

About the Author

Kristin Thiel is a writer and editor based in Portland, Oregon. Her first book with Cavendish Square was on Dorothy Hodgkin, a Nobel Prize–winning chemist and pioneer in X-ray crystallography, which was also the focus of Louis Pasteur's doctorate. She has worked on many of the books in the So, You Want to Be a… series (Aladdin/Beyond Words), which offers career guidance for kids. She was the lead writer on a report for her city about funding for high school dropout prevention. Thiel has judged YA book contests and helped start a Kids Voting USA affiliate. She has been a substitute teacher in grades K–12 and managed before-school and after-school literacy programs for AmeriCorps VISTA. She's had dental X-rays and walked through full-body scanners at the airport, but she's never broken a bone or swallowed a penny. She hopes to spend many more years not in a medical X-ray machine.